The Octopus and the Orangutan:
More True Tales of Animal Intrigue,
Intelligence, and Ingenuity

The Parrot's Lament:
And Other True Tales of Animal Intrigue,
Intelligence, and Ingenuity

The Future in Plain Sight:
Nine Clues to the Coming Instability

Silent Partners:
The Legacy of the Ape Language Experiments

Affluence and Discontent:
The Anatomy of Consumer Societies

The Alms Race:
The Impact of American Voluntary Aid Abroad

Apes, Men, and Language

THE WINDS

CLIMATE, WEATHER, AND THE DESTRUCTION OF CIVILIZATIONS

OF CHANGE

EUGENE LINDEN

SIMON & SCHUSTER

New York   London   Toronto   Sydney

SIMON & SCHUSTER
Rockefeller Center
1230 Avenue of the Americas
New York, NY 10020

SIMON & SCHUSTER and colophon are registered trademarks
of Simon & Schuster, Inc.

For information about special discounts for bulk purchases,
please contact Simon & Schuster Special Sales at
1-800-456-6798 or business@simonandschuster.com

Illustrations rendered by John Del Gaizo

Designed by Katy Riegel

Manufactured in the United States of America

1   3   5   7   9   10   8   6   4   2

Library of Congress Cataloging-in-Publication Data
Linden, Eugene.
The winds of change : climate, weather, and the
destruction of civilizations / Eugene Linden.
          p.   cm.
Includes index.
1. Climate changes. 2. Climate and civilization. 3. Weather.
4. Weather—Social aspects. I. Title
QC981.8.C5L567   2006
551.609'01—dc22   2005054434
ISBN-13: 978-0-684-86352-8
ISBN-10: 0-684-86352-9

To my father.
I only hope that I'm as
much of a delight
when I'm 86.

Contents

Contents

THE WINDS OF CHANGE

Preface

I'VE BEEN WRITING about nature and the environment since the early 1970s. I've watched as some environmental catastrophes have materialized, while with others it has been "Never mind." Lake Erie did not die as some predicted when I was a college student, and, as skeptics of global warming point out constantly, we have not begun a plunge into a new ice age, which was the prediction of some climatologists in the 1970s. On the other hand, even decades of insistent warnings could not prepare Americans for the actual horrors that Hurricane Katrina unleashed in August 2005. Turbo-charged by complacency, folly, and incompetence, Katrina destroyed a great city, transforming New Orleans into a septic stew of floating bodies, roaming gangs, disease, and toxic slime. The storm launched a wave of refugees not seen in the United States since the Dust Bowl, and the damage inflicted on crucial energy and transport infrastructure sent ripples throughout the economy.

If there is a message to take away from a look back at past predictions of potential calamity, it is that the risks of erring on the side of caution tend to be fewer than the costs of dismissing predicted threats out of hand. Alarms about Lake Erie mobilized people and governments to take action, and in proving doomsayers wrong, the cleanup also created billions of dollars in value as the lake area became a draw for real estate and recreation. While in the United States officials took action to clean up the air and water beginning in the 1970s, elsewhere in the world environmental threats such as deforestation and extinction are more critical today than they were thirty years ago.

Perhaps the most revealing aspect of this look backward to the early 1970s, however, is the threats that were not there, but which have since risen to prominence. Most prominent would be the possibility of climate change. In the late 1970s, climate specialists first started worrying about the possibility (and indeed a report submitted to President Jimmy Carter in 1979 was right on the money about noticeable changes in climate by the year 2000 if nothing was done to check emissions of greenhouse gases), but with the Iranian hostage crisis and stagflation dominating public concerns, the warning got little notice. The possibility that humans might be altering climate (performing a global experiment with us in the test tube, as some scientists put it) only became an issue in 1988, when Washington sweltered during an abnormal heat wave at the same time that Senator Timothy Wirth held hearings on the issue.

Assignments have taken me to both polar regions and out into the Gulf Stream in attempts to keep pace with the science of this unfolding story. Since 1988, public concerns about climate change have waxed and waned with the weather, but in the United States at least, climate change has not been a pressing issue for the public despite periodic alarms raised by scientists. I have more to say about this in Chapter 18, but at least part of the problem is that, for all practical purposes, the threat is unprecedented. In this respect, our attitudes toward climate change are a little like American attitudes toward terrorism before September 11, 2001, or the attitude toward tsunamis of a tourist visiting Phuket, Thailand, before December 26, 2004. With regard to climate, it's hard to imagine that we puny humans could affect something so all-encompassing as climate itself; it's hard to imagine what it would mean if climate started changing everywhere on earth; and today even those Americans who view climate change as a threat see it as an event that lies far off in the future.

After all, Americans suffer extremes of weather all the time without any long-term disruption of the economy. If climate change brings more extreme weather, the economy will absorb that too. Similar attitudes and such confidence might well have characterized the Akkadian priests in 2200 B.C., the rulers of the Old Kingdom in Egypt at that same time, the Mayan elite in A.D. 900, the Anasazi in the American Southwest, the Norse settlers in Greenland before A.D. 1350, and many other societies and civilizations which would discover that climate, as oceanographer Wallace Broecker puts it, "is an angry beast."

We humans have a difficult time estimating risk. We spend disproportionate energy worrying about statistically insignificant risks—e.g., being attacked by sharks—and yet are blasé about the risks of getting behind the wheel of a car. We are probably at our worst when estimating the risk of something, such as global climate change, that has not yet happened, or happened long ago.

For present-day Americans, the threat of climate change may be abstract because it is unprecedented, but the impact of climate change on other civilizations is not without precedent. And so perhaps the best way to understand the risks might be to look back at the ways in which climate change has affected successful civilizations in the past. This is an undertaking that has become possible only in the last decade (although visionary climate historians like H. H. Lamb began writing about the impact of climate on history in the 1960s), since prior to the 1990s, the picture of past climate was spotty, and in many cases the resolution too crude to link particular historical events with weather at a given time.

Until very recently, climate has been viewed as static. It was only in the mid nineteenth century that scientists discovered the wrenching changes of the ice ages, but even after that, the prevailing attitude was that the present 10,000-year warm period that gave rise to civilization was monotonously stable. So long as climate was viewed as predictable and stable, there was no pressing need to consider it a factor in the fate of civilizations. (It was the discovery that the warm periods between glacial eras tend to last 10,000 years that in part prompted fears of a new ice age in the mid-1970s.) A scientific paper published in 1997 entitled "Holocene Climate Less Stable Than Previously Thought" shows that the notion that our present climatic era has been boring and predictable persisted until very recently.

Nor have historians and archaeologists greeted climate historians with open arms. Those reconstructing the fate of ancient civilization already have to deal with a full plate of competing factors that could bring down a civilization without any deus ex machina like climate. An archaeologist who has been studying a matrix of trade relations, warfare, internal strife, and political intrigue is not going to drop everything when a paleoclimatologist says, "The weather did it."

In certain cases, however, the evidence is pretty compelling, not just in linking weather to a particular event, but also specific ways in which a changing climate may have undermined the legitimacy of rulers. In some cases, climate change fostered the spread of disease; in others, climate change might have set in motion a chain of events that led to migration and warfare. In one well-documented case, the cold alone made life untenable. The interplay of climate, politics and economies is complex, but there is evidence from the past that helps us sort this out.

I offer evidence that we disregard the role of climate in history at our peril. I've structured the book along the lines of a case. The opening section presents the prosecution's argument that climate change has either killed off or at least been an accomplice in the fall of several civilizations. It quickly runs through the various victims (and a notable evolutionary beneficiary), and also details the weapons and methods of this civilization killer.

The first chapter of Part Two explores how environmental factors, including climate, have fallen in and out of favor as forces affecting history. Subsequent chapters in this section offer a brief description of the gears of our climate system and then look into the forensics of climate history, describing and assessing the various proxies that paleoclimatologists use to reconstruct past weather. The section also suggests some big unanswered questions about past climate and how climate works, questions that have a bearing on our assessment of the present threat of climate change.

Part Three revisits the cases introduced in the opening section; it presents dissenting opinions and digs deeper into the implications of the proxy evidence. Part Four looks at El Niño as a force in history. Although that familiar event is not nearly as disruptive as other climate events of the more distant past, this regular cycle has had huge impacts on humanity at different times. Some historians argue that a series of El Niños in the late nineteenth century killed more people than the two world wars of the twentieth century combined. Moreover, there is a detailed record of El Niño's role in various historical events that reveals both the resiliency of the modern market economy as well as new vulnerabilities to changing climate.

In Part Five, we return to the present. The first chapter looks in detail at the peculiarities of the climate-change story as it has unfolded since the threat first surfaced. In the next chapter, I join a research expedition in the Gulf Stream to check the health of one of the vital organs of the global climate system. The final section draws on what happened in the past and what is happening in the present to develop a scenario of what we may face in the future.

We have an advantage over past civilizations that were blindsided by climate change. We can learn from their misfortunes.

PART ONE

Opening Arguments

1

A Matter of Emphasis

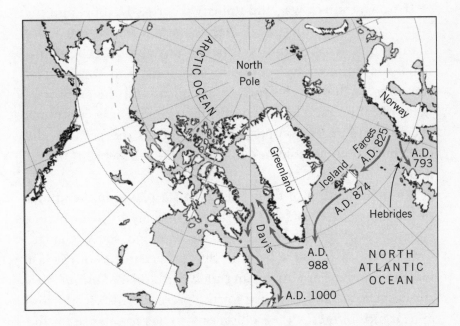

IN THE Francis Ford Coppola film *The Conversation*, the plot turns on the efforts of an audio technician who attempts to reconstitute a critical phrase caught by a clandestine bugging device after being uttered by a young and scared couple. At first, with the sound scratchy and grainy, it seems that the technician has revealed the plight of the couple as one says to the other, "He'd *kill* us if he had the chance." The technician does further filtering of the static in the message, and we see how subtle shifts in inflection give different meanings to the words. In the end, the technician discovers a sinister shift in meaning in the emphasis placed on one word: "He'd kill *us* if he had the chance." The emphasis on *us* reveals that it's the attractive young couple who are the killers. The alternate histories that derive from such subtle shifts came to mind recently as I pondered a new interpretation given to a much-studied, tumultuous period in European history: what Barbara Tuchman called "the calamitous 14th Century."

It was a time of war and upheaval across Europe, and the worst ravages of the Great Plague, which wiped out more than one-quarter of the continent's population. The continent was buffeted by economic, religious, and sociological fractures, including the end of feudalism and the empowerment of laborers. Amid the consideration of these enormous events, until recently one tiny footnote to the period—the end of the Norse colonies in Greenland—has been all but ignored.

An investigation of their fate reveals a player in the affairs of humanity that has been previously obscured by our preoccupation with grand events such as war and social change. This player casts a new light on the early centuries of the last millennium. Barbara Tuchman called her book *A Distant Mirror* because it had resonance for the present.* In fact, the disappearance of these Norse colonies also has resonance for the present. The calamity changed history, and the major factor in their disappearance will likely change history again.

The fourteenth century roughly marks the end of the Medieval Warm Period and the onset of the Little Ice Age, which began in earnest in Greenland in 1343 (the LIA began at different times in different parts of Europe), and continued on and off until the beginning of the nineteenth century. The Lit-

*Barbara Tuchman. *A Distant Mirror: The Calamitous 14th Century* (Knopf, 1978).

tle Ice Age was of more than meteorological concern. Its precedent, the Medieval Warm Period, which began in about A.D. 900, created wealth and prosperity throughout Europe. Perhaps no group in Europe was more favored by the mild weather than the Norse. The onset of the warm period coincides with the peak expansion of Viking influence.

The Vikings were Europe's first maritime power. In little more than a century, they established a presence over a great swath of the northern hemisphere, with outposts from Canada to Constantinople. They took over parts of France and the United Kingdom and threatened Rome. Such was their reputation and ferocity that at the height of Viking power a Norsewoman named Gudrid, accompanied by a few of her countrymen, could journey on land from Denmark to Rome and back unmolested, a feat no Roman would have undertaken without an army, even at the peak of its empire. Then, after just a century or so of dominance, Viking power began to fade. Its colonists became so assimilated into various European cultures that when William the Conqueror defeated King Harold Godwinson of England at the Battle of Hastings in 1066, an observing geneticist would have seen it as a battle of Norse versus Norse.

Over the years, historians have looked at the grainy tape of the Viking era as revealed in the sagas, accounts of monks, and the meager artifacts recovered from the earth. Some historians have argued that one need look no further than politics and technology to understand the Vikings' rise and fall. At home in Norway, for instance, Harald the Fairhaired consolidated power in the late ninth century, and those members of the nobility who did not like paying taxes or taking orders decided to put their warrior skills to good use and establish themselves in France, the British Isles, and anywhere that offered an opportunity to a bright, entrepreneurial, and ruthless young man. As the celebrated Norwegian anthropologist Vilh-jalmur Stefansson, put it in his introduction to *The Vinland Voyages,* published by the American Geographical Society in 1930, the Viking expansion was perhaps "the only large-scale migration in history where the nobility moved out and the peasantry stayed home."*

The Vikings sailed to power abetted by technological advance. By the middle of the eighth century, they had figured out how to use overlapping boards and waterproof pitch to build a ship that could articulate with the waves, and how to secure and distribute the weight of the huge masts needed to hold the sails that powered the 75-foot oceangoing vessels. This gave the Norsemen a huge advantage over other mariners.

In recent years, a number of historians have been looking at the grainy tape of this era in a new light. The chronology of the rise and fall of the Vikings has not changed, nor have the signal events. What has changed is that oceanographers, glaciologists, geophysicists, and other scientists have teased a picture of the context of these events, revealing an agent working in the background. This agent first smoothed the way for the Viking Age like a benevolent God, and then, suddenly, took it all away.

The agent was climate, something so all-embracing that we tend to ignore it. After all, climate is context, the playing field, the given. How can climate be a factor in human events since European cultures have presumably adapted to European climate?

But climate changes, and when it does, it favors some and penalizes others. This is what happened during the Viking Age. The ways in which it favored the northern folk were subtle, while the harsh events that penalized their descendants were not. The interplay of climate and the Vikings stands out

*Matthias Thordarson. *The Vinland Voyages.* American Geographical Society Research Series No. 18 (1930).

because here was a case in which climate played a crucial role in shaping the world we live in today.

At first, climate favored the northern peoples. The halcyon period, referred to as the Medieval Warming or the Climate Optimum, saw significantly warmer temperatures through much of the North Atlantic. Farmers grew grapes in the United Kingdom, and population and trade expanded throughout Europe. Significantly for the New World, this warming period had a dramatic effect on places like the Faroe Islands, Iceland, and Greenland, which are located at the northern margins of cultivable land and thus benefited disproportionately. In parts of southwestern Greenland, the growing season lengthened to allow the planting of crops such as barley, while lush meadows supported goats, sheep, and cattle.

The unusual warm period created new options for Viking adventurers. While Hrolf the Ganger fought his way up the Seine (eventually prompting the French King Charles III to offer him all of Normandy if he would stop his incessant raids on Paris), other Norsemen headed westward. Although some historians speculate that the Norse were capable of primitive celestial navigation, the best guess is that they navigated by island hopping on a grand scale. Gunnar Eggertsson, who painstakingly replicated a ninth-century Viking vessel, and then in 2000 retraced Leif Eriksson's expedition to North America, argues that the Vikings liked to keep land in sight. "On a clear day, you can see the Shetlands within a day's sail from Europe," he says, "and the same goes for the Faroes from the Shetlands." Iceland is a bit farther away from the Faroes—252 miles—but, says Eggertsson, it's possible to see the clouds above Iceland, if you venture northeast from the Faroes in exceptionally clear weather.

The Vikings were established on the Faroes for decades before they made it to Iceland. It is almost certain that a Viking venturing offshore caught a glimpse of the island, confirming the tales about travels there by Irish hermits who preceded the Vikings. Legend has it that the hermits got to Iceland by hopping into their little coracles and following geese.

The first Norse who ventured to Iceland starting in 874 could scarcely believe their good fortune. Trees were plentiful (the island was about 30 percent forested then, as opposed to the 1.3 percent forest cover that remains today), though the birch and willows were not as grand as the hardwoods in Europe. Seemingly endless grazing lands beckoned, and the island was uninhabited so that, at least at first, one could seize land without any tiresome battles with natives or other Norse. Hot springs supplied infinite amounts of warm water (there is evidence that Vikings were poaching themselves in hot tubs 1,000 years before Californians). Within eighty years of their first settlement, as many as 50,000 Norse lived in Iceland. Since the largest Viking boat might hold about 70 people, it seems clear that there was heavy traffic across the North Atlantic.

At first, Iceland was a Utopia. The noblemen and chieftains set up a libertarian paradise where landholders could do their own thing so long as it did not disadvantage their neighbors. Disputes and general rules were worked out at summer meetings of the *althing,* the world's first truly democratic parliament.

The Norse even demonstrated sensitivity to environmental issues. One early edict that came out of the *althing* banned the grazing of livestock after landowners discovered that sharp-hooved animals such as sheep and goats were rapidly destroying the delicate ground cover. Some students of Icelandic history believe that Iceland's laissez-faire system coalesced into a state because the concerned chiefs needed a way to come to grips with environmental degradation.

There was just one catch to this idyllic life.

By the year 982, people were crowding each other, and the

newly settled Vikings had not entirely put behind them their gusto for bloodshed. Settled and democratic they may have been, but they were also some of the toughest warriors who ever lived. The Vikings had a code that gave high honor to those who could get off a quip when mortally wounded. One of the most memorable lines from Viking sagas memorializes the wit of one hefty warrior who, when he had his stomach sliced open in a fight with natives in the New World, looked at his exposed innards and gasped, "Fat paunch that was . . ."

These bloody duels had reverberations far beyond Iceland. As Stefansson drolly put it in 1930, "The eventual discovery of North America hangs upon a fashionable practice of the day, that of man-killing, which, like cocktail shaking in the later America, was against the law but was indulged by the best people." He is referring to a few recidivist killers like Erik Thorvaldsson, Eric the Red, who overtaxed even the Norse tolerance for conflict and suffered the fate of being exiled more than once. The first time, he was forced to relocate to Iceland's west coast, and then he was voted off the island altogether.

Where to go? From Iceland's far west coast, Greenland—172 miles away—looms as a far-off and intriguing white mass. Eric made the jump across, and then followed the coast southward around the southern tip and up the western shore of the giant island. It took him a while to find a patch of green, but he eventually found a lush fjord head on the west coast which, during those clement times, had better land for growing barley than Iceland, as well as birch and willow trees, and meadows to support livestock. In the year 981, he put down stakes at a place referred to in the Greenland sagas as Brattahlid, which is located on a fjord on Greenland's west coast.

He promptly set about luring other settlers. The incongruous name of the island, which is dominated by the only ice cap in the northern hemisphere, comes from these attempts. As Stefansson put it, Eric the Red displayed "a genius for advertising that made him prophetically American." (There are various stories how Iceland, which is mostly green, got its reciprocally incongruous name. Most likely, it was because the route to Iceland carried Vikings past the coldest part of the coast, which sits in the shadow of one of Iceland's largest glaciers.)

The Vikings explored the western coast of Greenland, preceding by about a hundred years the Thule Eskimos, who were expanding eastward from the west. As they explored northward in search of walrus meat and ivory, the Norse got their first glimpses of North America. From the top of Disko Bay, 350 miles north of the Western Settlement, one can see across the Davis Strait to Baffin Island. From there, it's possible to follow the currents and move south along the Labrador and Newfoundland coasts without losing sight of land.

Having heard accounts of early forays into this New World (Bjarni, the son of a good friend of Eric's, had earlier ventured close but never landed), Eric's son Leif went looking in the year 1000 to see what he might find. He was accompanied by a wealthy and clever merchant named Thorfinn Karlsefni and Thorfinn's wife, Gudrid. Leif or Thorfinn established a beachhead, very likely in L'Anse aux Meadows in Newfoundland, a Viking site uncovered in 1960 by Helge and Anne Ingstad. Three years after first landing in the New World, Gudrid gave birth to a son, the first, and quite possibly the last, Norse baby born in North America. Given the rapidity with which the Norse populated Iceland, this birth dearth is a mystery that hovers over the Vinland experience.

Almost certainly, other Norse traders ventured to North America during the 350-year span of the Western Settlement in Greenland. It's entirely possible that there were scattered Norse settlements on the continent and that other babies were born. But no major presence other than L'Anse aux Meadows has yet been uncovered, and, manifestly, the Norse ventures in the New World petered out.

How could it be that a people, who in just a few decades could fully populate a windswept, marginally habitable island like Iceland, could fail to establish themselves on a continent with the tallest trees, richest soils, and most abundant wildlife in the world? There are many reasons, of course, not the least of which was the presence of natives who may have started out friendly, but, after a few encounters with the violent, undiplomatic Norse, quickly became hostile. John Steinberg, a UCLA-based archaeologist, notes that the Vikings were opportunists—"If you were asleep they would kill you and if you were awake they would trade with you." At different times, Leif and Thorfinn's men did both, and eventually, because the former practice did not sit well with the Native Americans, they were forced to retreat to Greenland and Iceland.

It's also possible that the Norse never saw the riches that lay to the south of L'Anse aux Meadows, and failed to realize that they had discovered a new world. Knowing what we do now about the vast wealth of the Americas, this might seem strange. The arriving Norse, however, would have had no way to know the scale of the landmass. Nor would they be the first to overlook a planetary-scale geographical feature. Abel Tasman discovered Tasmania in 1642, but, fighting bad weather, never noticed the giant continent of Australia that lay to the north and sailed clear around it.

There are many possible explanations for the failure of the Viking experiment in the New World. For one thing, Vinland was at the extreme outer edge of the Viking expansion. To draw on labor, Norse entrepreneurs would have had to recruit from Greenland, which never had more than a few thousand Norse, or lure settlers from Iceland or Norway.

In this respect, the Norse presence in the New World might

be seen as analogous to Alexander the Great's adventures into India. Although India probably offered more riches than any other lands Alexander entered, his empire was so attenuated and his troops so homesick at that point that he could not consolidate power. The Norse in the year 1000 had neither the numbers nor the supply lines to feed large numbers of people into North America.

Birgitta Wallace, a Canadian archaeologist who has spent much of her career studying the Norse presence on the continent, also notes that they never lost their orientation toward Norway and Europe. The farther they ventured westward from their cultural center (ventures east never brought them too far from European influence), the weaker the hold of wherever they were.

There is also the question of the motivations of the original explorers. They were not driven to the New World by religious or political persecution (Harald was more an irritant than a true despot). Nor were they conquering new lands for the pope or the king. Rather, like Thorfinn, many were looking for opportunities for trade and plunder—the literal meaning of the phrase to "go a viking." Because they were opportunists, if the costs of working in one place got too high—e.g., hostile natives in Vinland showering you with arrows—it was relatively easy to pick up stakes and find other opportunities.

Still, there was one factor above and beyond these political, social, and economic considerations that for practical purposes eliminated the possibility that the Norse would colonize North America. The weather turned awful. Seafarers had to deal with increasingly frequent intrusions of sea ice as early as the mid 1100s, a cold period that offered a preview of the Little Ice Age and the famines to come.

Starting between 1343 and 1345, Greenland suffered through ten cold years, culminating in the worst winter in five hundred years in 1355. This led to the collapse of the western colony. As Thomas McGovern, an archaeologist and professor of ancient history at Hunter College put it, "A northern economy is equipped to withstand the occasional bad winter, but not year after year of extreme cold."

One hundred years later, the more southerly Eastern Settlement in Greenland succumbed as well. Without these settlements, Norse seafarers had no place to stop and resupply. Moreover, an increase in sea ice made the trip ever more dangerous (and would have dissuaded mariners from sailing directly across to the New World, even had the Norse developed open-water navigation). Norse ventures in the New World were no longer feasible.

How different the world might be today if the Medieval Warm Period had continued and those Greenland settlements had endured.

If that had happened, the Norse colonies might have become better established and other explorers would have ventured to the New World. Norse explorers would have eventually realized the scale of the riches to the south, lands brimming with game, rich soils, and gigantic, champion primeval hardwood forests that seemingly extended to infinity. Tall straight trees were coveted for masts, and in Europe masts were the key to maritime power. For a Norse nobleman seeking to establish his own power base, the sight of such forests and lands would have been well-nigh irresistible.

Over the years, some Norse adventurer would have found a way to deal with the Native Americans, more likely through accommodation than conquest (a good start would have been to find a more diplomatic term for aboriginals than the Norse word, Skraelings, which means "wretched people" and was used to refer to both Inuit and subarctic Native Americans). The Norse of that time were not carriers of the infectious diseases that became the weapons of biological warfare that five hundred years later tipped the balance in favor of British and

Spanish invaders. (McGovern notes that support for this argument comes from the fact that a smallpox epidemic killed one-third of the population on both Iceland and Greenland—something that would not have happened if the Scandinavians had prior exposure to the disease). A nation that could put 50,000 people on a harsh uninhabited island in a matter of decades could also rapidly colonize the richest lands on earth. Had the weather remained good, it's entirely possible that some parts of what is now Canada and the eastern United States would today be speaking Norse, not English.

This did not happen.

Instead, life for the Norse quickly turned into a nightmare. Over two decades, Thomas McGovern, has pieced together a picture of just how bad life became as he unraveled a six-hundred-year-old mystery of death and disappearance in one particular farmstead, catalogued as site V54, which was uncovered at the location of the Western Settlement in Greenland. Beneath the sands, well preserved in the arctic environment, were the remnants of a sod-and-wood house. Excavation of the farm at Nipaitsoq began in 1976, led by Claus Andreasen and Jens Rosing of the Greenland Museum. McGovern came later to tease a story out of their careful excavation. It's not a simple story of climate turning cold and people dying. McGovern points out that as early as 1911 famed Norwegian explorer Fridjof Nansen speculated about the role of the cold in the demise of the Greenland colonies. Rather, it's a story in which environmental degradation left the Norse vulnerable when climate turned bad. "Drawing down natural capital is setting yourself up for environmental change," says McGovern.

The ancient Greenland Norse were a curious mixture of opportunistic and hidebound in their adaptations to new lands. For their housing, they adapted the basic model used in Iceland: a structure built from turf, stone, and wood if it was available. They took pains to conserve heat, digging their buildings into the earth and covering the floor with brush and twigs. The farmhouses contained several relatively small rooms designed to hold animals and people, and to conserve heat during the cold winters.

For food, they relied on hunting and gathering as well as farming. They would harvest bird's eggs and kill harp seals when they appeared during the spring. They also imported large hunting dogs from Europe to hunt caribou. The Norse used the summer months to grow grains for livestock—cattle, sheep, and goats—and then relied on meat and milk to get them through the winter.

When McGovern started on this puzzle in the late 1970s, he had relatively skimpy and imprecise scientific evidence that weather played a life-and-death role in the viability of this settlement. In a 1982 paper, "The Lost Norse Colony of Greenland," he refers to pollen data (revealing of changes in vegetation) and early ice cores taken from Camp Century on the Greenland Ice Sheet. Still, even then he suspected that the Little Ice Age had something to do with the demise of the farmstead. He theorized that increased sea ice cut off the Norse from their supply lines to the east, while the long snowy winters forced the livestock to be kept in cramped byres far longer than normal, fatally weakening the animals. Instead of delivering their calves on rich meadows, near-starving animals gave birth in the filthy byres. The late springs also delayed the arrival of the harp seals and exhausted the settlers' supplies of seal, caribou meat, and cheese.

At the time McGovern published his paper, few archaeologists acknowledged climate as a force in society. Moreover, the proxies then available to reconstruct climate history were unreliable or had only crude resolution and were not particularly useful in determining what a general pattern of changing climate meant for a specific area. Written accounts were skimpy too. McGovern notes that at that time scientists interpreting the data did not have confidence assigning dates on timescales less than a hundred years. All in all, this was not the type of evidence that would give a historian confidence when linking an actual event in a specific place with the weather in a given season or year. Yet that's what McGovern did. Then, in the mid-1990s, McGovern's argument got an enormous boost from a project that had nothing to do with archaeology.

These were the findings of the Greenland Ice Sheet Project. Launched in the late 1980s, GISP2 convened researchers from a variety of institutions and scientific disciplines to gather ice cores from the two-mile-thick ice sheet. It was hoped that GISP2 would either confirm or contradict dramatic, but ambiguous, signals of rapid climate change collected by Danish scientist Willi Dansgaard in 1983 in a different part of Greenland. The climatic timeline gleaned from these cores proved to be a bonanza for McGovern.

From the ice cores, oceanographers, chemists, paleoclimatologists, and other scientists have been able to reconstruct accurate variations in past climate going back 110,000 years. The climate changes documented by the cores took place right on V54's doorstep. McGovern, however, did not have data from this giant project until Lisa Barlow of the University of Colorado began publishing studies of the cores taken from medieval times. Also during the mid-1990s, data from other attempts to reconstruct past climate began to become available. Studies by Woods Hole Oceanographic Institution geophysicist Lloyd Keigwin of the skeletons of foraminifera—popcornlike creatures from the Sargasso Sea—helped fill out the picture developed earlier from the record of tree rings, the evidence of historical accounts, sagas, and even paintings.

Forensic analysis of insect remains helped establish dates when the weather changed and provided proxy evidence for events unfolding inside the house that McGovern and colleagues were studying. With each study, with each new piece of evidence, the story of life in this one house in Greenland in the years between 1345 and 1355 became more detailed, more convincing, and more horrifying.

McGovern's legend of the Western Settlement's fall is built upon the order in which bones, insect remains, and other materials were entombed in the nearly sterile glacial sands of the farmstead. By analyzing and dating the remains found in different layers, the archaeologists could make informed guesses about what happened and where. They could determine the location of such rooms as the larder, the kitchen, and the sleeping quarters, as well as what was going on in those rooms.

McGovern was struck by the order in which the bones of the animals were uncovered. In the terminal layer—from 1355, the last year with evidence of any habitation—they found cow hooves in the larder, suggesting that the Norse had eaten their dairy animals, something they would only do in extreme circumstances, and only the truly desperate would gnaw at the hooves for sustenance. They also found dog bones inside the house, not in front. Moreover, the dog bones had tooth and cut marks on them, evidence that the animals had been killed and eaten. Among the bones in the terminal layer were the foot bones of ptarmigan and hare. "That's starvation food for the Inuit," says McGovern, "because there is so little fat."

Another clue to the fate of the farmstead came from insect remains. By identifying the tiny shells of insects entombed in a given layer, archaeologists can make informed guesses about the weather inside and outside a structure, as well as the health, comfort, and diet of a household. An interpretation of these microscopic clues came from Paul Buckland, an environmental archaeologist at Sheffield University, and from Peter Skidmore, who does forensic studies of fly remains for Scotland Yard and otherwise spends his time as an entomologist at Sheffield.

The insects filled in much of the story. Evidence of happier times at V54 came from deeper levels, which contained the remains of *Teleomarina flavipes,* a fly typically found in heated buildings and warm climates. Toward the frigid endgame of this colony, these fly remains give way to a cold-tolerant species that feasts on carrion. At first, these carrion fly remains are found in the larder, but in the last year they show up in the sleeping quarters. When he saw this, Skidmore remarked to McGovern, "if this was a crime scene, we'd be looking for a body." But there were no bodies.

Over the years, McGovern has put these findings together into a narrative of slow collapse and starvation. Starting between 1343 and 1345, the weather turned sharply colder. The Norse way of life was designed to cope with one bad year, but not ten to twelve bad years in a row. Short cold summers gave the Norse no opportunity to rebuild their flocks and grain supplies. They shifted their diet from 80 percent mutton to 80 percent seafood, but seafood, which ordinarily could have sustained the population, became harder to get as the ice closed in more frequently. The Norse ate the sickly calves born in the byres, and then the livestock itself, a desperate measure since it eliminated the possibility of storing milk and skyr (a yogurtlike cheese) for the next winter.

McGovern thinks the end came quickly during the dreadful year of 1355. The doors were found lying flat, a significant clue, says McGovern, because the Norse always took their doors with them during an orderly departure from a settlement. It is possible they somehow evacuated, but why then would Skidmore find carrion flies in the bedroom. It's also possible that neighbors, Norse or Inuit, subsequently found bodies and buried them.

Many historians hold that it is simplistic to say that the residents of V54 were victims of climate change. While climate penalized the Norse who clung to their traditional ways, Mc-

Govern himself points out that the Inuit flourished during this same period. The Norse could have survived the bad weather too if they had learned from the Inuit, who love it when the weather turns frigid because it gives them an ice platform from which to hunt ringed seals with harpoons when the mammals surface at breathing holes in the sea ice. There is no evidence that the Norse hunted ringed seals. McGovern has said that the Christian Norse likely regarded the shamanistic Inuit as unenlightened and beneath them. Birgita Wallace attributes this fatal blind spot to ethnocentrism. Other factors cited for the decline of the Greenland colonies include economic competition, as African ivory emerged to compete with the walrus ivory that had been the principal Norse export.

The more southerly Eastern Settlement held on for perhaps another hundred years. Some date the collapse of this colony to 1500, based on the finding of a cap that became fashionable about that time. McGovern, however, believes that this "cap" is a misidentified woman's headdress that was worn throughout the period. Moreover, he notes that pollen data shows that trees started coming back after 1450, something that would not have happened with people in the area cutting timber for firewood and building materials.

Today, it's hard to argue with McGovern's now confident assertion that "the Little Ice Age proved absolutely fatal to the Greenland colonies." McGovern believes that the Vikings had one window in time in which they might have colonized North America: in the years around A.D. 1000. The reason they did not exploit that opportunity may have had nothing to do with climate, but it was climate that ultimately closed the window of opportunity for the Norse in the New World.

In this climatic version of *The Conversation*, a picture emerges of a culprit—changing weather—all but ignored when archaeologists first discovered site V54 in 1976. This might be nothing more than a historical curiosity were there not increasing evidence that the end of the Norse presence in Greenland and the possibility of colonizing the New World was not an isolated instance. Emerging evidence suggests that climate may well be a serial killer of colonies and even civilizations. This culprit may have had a role in the collapse of the Akkadian, Moche, Mayan, Anasazi, and other civilizations.

Many historians, archaeologists, and anthropologists dispute the role of climate as a factor in history. John Steinberg, the UCLA archaeologist, put this reaction succinctly: "Most archaeologists are anthropologists at heart, and most anthropologists hate the assertion that humans are not in control of their destiny."

For every example of a historical collapse coincident with a dramatic shift in climate, there is an archaeologist or historian who will argue that technological, cultural, political, or economic factors were more important. In the course of this book, I will try to fairly represent these counterarguments. Climate history is still a very young field.

Before GISP2 and other recent breakthroughs that allow reconstruction of past climates with resolution to a specific year, those who tried to link climate and historical events, pioneers such as the late British historian H. H. Lamb, had to base their arguments on snippets of evidence—ancient accounts of harvests, the levels of the Nile, tree rings, and, here and there, the analysis of lake-bed sediments. In places with rich records, such reconstructions were possible, but for much of the world, the image of past climate was smudgy and incomplete. As recently as 175 years ago, even the notion that great ice sheets reshaped the mountains and valleys of Europe was considered preposterous.

Climate may be a newcomer as a factor in the fates of civilizations, but it is becoming clear that changes in climate have pushed some societies beyond their capacity to adapt. The Norse could have adopted Inuit hunting methods and survived the Little Ice Age. Similarly, the Anasazi of prehistoric New Mexico could have changed their ways and survived the drought that brought about their disappearance. But neither did.

Maybe the Norse were homesick; maybe the appearance of African ivory took away their economic raison d'être (by depriving them of income from walrus ivory); maybe a Europe depopulated by the plague offered available and better farmlands. Perhaps all of these factors played a role, but the simple fact is that the Greenland colonies prospered during the warm years and became uninhabitable by agrarian people during the cold years. Whether climate is the principal culprit or accessory is a question I will try to sort out.

Climate has used different weapons to bring down the mighty. Most of the time, it is drought that upsets the balance of power. As Richard Alley, one of the lead scientists in GISP2 wrote in *The Two-Mile Time Machine,* "Through history it has been changes in precipitation, not temperature, that has brought down civilizations."* One of the rare exceptions, however, may be the case of the Norse colonies. So, the question remains, what was it that brought about the short cold summers and intensely cold winters that afflicted Greenland, Europe, and many other parts of the world on and off for roughly five centuries?

If we look broadly at the Little Ice Age as a product of some long-term natural cycle, not likely to return for several hundred years, then we may comfortably treat it as a historical curiosity, revealing the ways in which changing climate can buffet human societies, to be sure, but relevant only as a cau-

*Richard B. Alley. *The Two-Mile Time Machine: Ice Cores, Abrupt Climate Change, and Our Future* (Princeton University Press, 2000).

tionary tale. A closer look, however, reveals other factors, some of them unknown until just the past few years, that have a compelling and urgent message for the present. These factors have to do with the *how* of the Little Ice Age.

Climate scientists have tried to come up with explanations for the onset of the Little Ice Age. Some cite a change in sunspot activity—called the Maunder Minimum by astronomers—that reduced solar radiation by about 1 percent during the coldest part of the period starting in 1615. Others have pointed to a spate of volcanic eruptions that threw a pall of dust into the atmosphere and blocked some of the incoming sunlight. Astrid Ogilvie, of the Institute of Alpine and Arctic Research at the University of Colorado, has argued that the whole thing is a misnomer—that the Little Ice Age was more a period of climatic volatility than of persistent cooling. Paul Mayewski, now at the University of Maine, asserts that this event was the latest expression of a regular cycle of cooling and drying that dates back tens of thousands of years.

Since the Little Ice Age began at a time when there were not many people and the use of fossil fuels was in its infancy, it is not likely that humans had anything to do with starting the LIA (though recently some scientists have argued humans have long impacted climate through clearing and agriculture). The cold period probably ended for natural reasons as well (that is, if it ended, an issue that will be dealt with later). Before relegating the event to historical curiosity, it's worth considering a couple of coincidences and questions.

The first is that even if the Little Ice Age represented a minor perturbation of climate, it seemed to have outsized effects on European history. Its effects on the Norse represent a recent addition to a large body of literature on the impacts of the event. The second is that the cooling may have started at different times in different parts of Europe, but when it did come, temperatures dropped fairly rapidly, and new evidence indicates that wintertime temperatures during the depths of the LIA dropped to extremes. Moreover, one of the astounding results coming from GIS2P was that deeper in the past, climate made many large, sudden shifts from warm to cold and from cold to warm. What caused these shifts, and did this same factor or factors play a role in the Little Ice Age? Framed this way, the *how* of the Little Ice Age becomes relevant to the present because the answer might help tell us what we are in for in the coming decades.

One possibility is that the event involved changes in ocean circulation. This suggestion was proffered by the creative paleoclimatologist named Wallace Broecker of Columbia University's Lamont-Doherty Earth Observatory in Palisades, New York. Broecker had earlier gained fame for first describing in 1987 what he called the "Global Ocean Conveyor Belt," a ribbon of giant ocean currents that flows around the planet, distributing heat as it does so. Part of this flow, known familiarly as the Gulf Stream, carries warm water northward in the Atlantic Ocean, tempering the climate of Europe and North America as it does so. The system moves an enormous amount of water. At the point at which this current dives to form deep ocean water in the North Atlantic, the conveyor moves more than the amount of water of one hundred Amazons. Estimates of the amount of heat transmitted to Europe and other northern areas are on the order of 1,000 times the generating capacity of the United States. Stefan Rahmstorf, an oceanographer at Germany's Potsdam Institute for Climate Impact Research, estimates that the amount of heat in the system is equivalent to the output of a million power plants.

The habitability of northern Europe is to some degree (just how much is a subject of debate) a gift of the oceans. Every schoolkid knows that, but Broecker discovered that this is a gift that does not always keep on giving. The Gulf Stream's warm embrace can be withdrawn, and quite suddenly.

The engine that pulls this current northward is the weight of the dense salty water of the northern part of the current. As it moves northward, evaporation makes the water saltier and heavier, and when lobes of the current pass over an underwater sill in the far north, the weight of the salt water carries it down into the depths, in effect, pulling the current behind it. Because it involves salt and heat, the system is commonly referred to as thermohaline circulation, or THC. I find it nothing less than astonishing that this colossal heat engine is propelled by something as subtle as the difference in water densities between a current and the surrounding ocean waters. For all its energy, this conveyor has a delicate side.

What if, Broecker wondered, melting glacial ice and the increased rainfall that comes during a warm period poured a lot of fresh water into the North Atlantic where the remnants of the Gulf Stream dive to form what is called "deep water." Would the fresh water mix with the salt water of the current, making it fresher and lighter? In that case, instead of diving, the current would either peter out or shift. Either way, the northern countries would be deprived of an enormous source of heat. As the influence of the Gulf Stream lessened in the far north, sea ice could expand in the winter. Sea ice both reflects sunlight back into space and traps underneath it ocean heat that would otherwise warm the surrounding air. With changes in the Gulf Stream and sea ice, climate could shift rapidly from temperate and maritime to arctic.

Something like this took place just as the world was warming after the end of the last glacial period about 12,800 years ago and in just a few years plunged the world into a 1,300–year deep freeze. The discovery of the rapidity and extreme amplitude of this change alerted the climate community that climate is not the lumbering sloth many had thought, but more like the "angry beast" that Broecker describes.

During the past decade or so, the community of oceanographers and paleoclimatologists have been probing the past with ever greater urgency to determine whether the shutdowns of ocean circulation that took place in the glacial era have continued into the current warm period that incubated civilization and saw the extraordinary rise in human numbers. As Broecker has written, "If deep watermass changes have occurred in the past several hundred years, then the ocean is not in a steady state . . ." and this calls into question all assumptions about future changes in climate.

There is increasing evidence of sudden shifts during historical times. Lonnie Thompson, a specialist on tropical glaciers at Ohio State offers evidence from around the world of a virtual overnight plunge in temperatures 5,200 years ago, and other scientists have uncovered evidence of the sudden onset of extreme drought and storminess that have destroyed many ancient civilizations. Were these abrupt events brought about by changes in ocean circulation, or by some other factor such as changes in storm tracks, episodes of volcanism? What has become clear, as Penn State geochemist Richard Alley puts it, is that while we've tended to comfort ourselves by thinking that climate change is like turning a dial, the reality is that shifts in climate are more like flicking a switch.

Some of the elements that probably produced past sudden changes in ocean circulation—dramatic warming, the melting of glaciers and increased precipitation, and the resulting freshening of the North Atlantic—are in place today. The 1990s were the warmest decade in the past 1,000 years, according to Michael Mann of the National Climatic Data Center (some dispute his methodology, but the data seem to support his contention). Moreover, a slowdown in thermohaline circulation may have already begun. A study of changes in water density published in *Nature* estimated as much as a 25 percent decline in the vigor of the circulation since the 1950s. The great engines of the THC are two sites in the northern seas where surface water sinks to form deep ocean water, pulling warm water north behind it. There has been no deep sinking observed at either of those sites in recent years. One shut down in 1982, the other in 1998, and no one can say whether this work stoppage is natural or the beginning of a major climate swing.

Stefan Rahmstorf, a climate modeler at the Potsdam Institute for Climate Impact Research, points out that this estimate is the result of indirect measurements and that oceanographers are just beginning to measure flow systematically. He says that it could take up to thirty years to develop a meaningful picture of what is going on, and notes that by then climate might already be changing dramatically. The ocean conveyor is a neat image, but in reality the current is not so packaged and coherent.

Thus, in the past ten years, a new, but only partially resolved, picture of how climate changes has begun to emerge. When the notion of rapid climate change began to capture wide attention among geophysicists and other climate specialists in the early 1990s, the idea that climate could change on a dime was a radical notion. Now the idea is moving rapidly to center stage.

The answer to the two crucial questions—what climate change means and what it might bring—has changed dramatically as our ability to decode evidence and filter out static has improved over the years. Moreover, the possible answers that have emerged have dramatically increased the importance of understanding the dynamics and historical effects of the Little Ice Age, as well as the meaning of other times when climate has moved from the background to the foreground and affected the course of history.

In this book, I will act as a profiler of this suspected serial killer of colonies and civilizations (although I should stress that this particular serial killer is a product of geophysical forces, and has no particular animus to humanity, nor any moral dimension). Profiling this suspect means evaluating the current state of knowledge about the ways in which climate changes, as well as exploring the ingenious new ways the various scientific disciplines have devised to reconstruct past changes in climate. As this grainy tape of the past becomes clearer, it has also become possible to revisit the demise of civilizations through time and across the globe as well as the interplay of climate, economics, disease, and politics that has spelled the difference between survival or succumbing.

2

The Deep Past: Climate as Creator

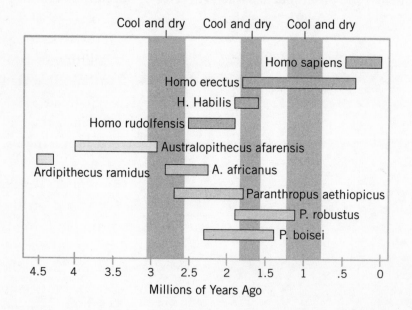

THE INTERACTION BETWEEN climate and ecosystems is a complicated dance. Grand events affect climate, and climate changes the risks and rewards for all living things. On the vast timescale of plate tectonics, for instance, events such as the continents settling into their current configuration coincided with the first major efflorescence of our ape ancestors, according to David Begun of the University of Toronto. Without the evolutionary experiments stimulated by the movement of earth's surface plates, there would have been no humans millions of years later. In evolutionary terms then, we owe a shout out to the continent of Antarctica. Far removed from the savannahs and rainforests of Africa, the separation and glaciation of Antarctica set the stage for a general cooling of the planet that threw our ancestral forms into an evolutionary ferment.

Antarctica separated from the rest of the ancient supercontinent Gondwanaland and reached its current location more than 30 million years ago. Its ice sheets had formed a few million years earlier, turning what had been a temperate rainforest and home for dinosaurs into a frigid wilderness of ice. Antarctica's initial glaciation may have been in response to climate changes brought about by the vertical uplifting of land masses and a period of volcanism, which partially blocked sunlight, but as more and more open water separated South America and the Antarctic Peninsula, ocean currents and winds could circle the globe, creating an ice box in Antarctica, thermally isolated from the rest of the world.

The open water allowed winds to sweep around the world, which is why the Southern Ocean is the roughest on the planet. For nine months of the year, the stratospheric winds circling above these waters form what is called a polar vortex, an atmospheric barrier against weather systems from the lower latitudes. To a large degree, Antarctica makes its own cold weather. Its separation helped set the table for a cooler world, and a more volatile climate.

Tectonic events began receding in importance about 20 million years ago as the continents settled into their present form. This was about the time of the first major efflorescence of our ape ancestors as they split off from monkeys. Even with a rather fixed map of oceans and land, however, events such as the uplift of mountain ranges and variations in sea level reworked climate. According to Begun, as wind and ocean currents responded to rising mountains and changing ice cover, the cooling and drying that resulted in Europe and Africa gave rise to new species and also killed off dozens of ape genera. He points out that the rising and falling of sea levels as ice sheets waxed and waned alternately isolated and reunited various ape lines, fostering a rainbow of genetic experiments in response to climate challenges, at least some of them involving increased brain power.

These experiments continued with the advent of the Pleistocene ice ages about three million years ago. The ice ages rendered irrelevant the arduously perfected strategies of creatures adapted to warmer periods and presented an evolutionary opportunity for those who adapted to colder weather and who could change their habits as climate changed. Some of the many hominin lines that comprise our ancestry fell into this latter group. Even as climate change led to the extinction of many species, it created the circumstances for the emergence of bigger-brained hominins (the term refers to those Hominidae more closely related to modern humans than to apes) at critical times during the several-million-year period of our ancestor's divergence from the rest of the apes and the arrival of *Homo sapiens sapiens*.

At least this is the argument advanced by a number of scientists, perhaps most prominently by Richard Potts, who directs the Human Origins Program at the Smithsonian Institution. Using data on ancient climate drawn from the analysis of seabed sediments, ice cores, and other records, Potts shows that periods of rapid speciation of our direct and collateral ancestors coincided with periods of great climate variability in Africa, and that climate variability favored the most adaptive and materially advanced of the hominins during these periods.

The onset of the most recent ice ages some 2.5 million years ago also initiated a general cooling and drying in Africa.

The trend was punctuated with great swings of climate, periods during which climate would remain unstable for 100,000 years or more. Piecing together climate records with fossil remains, Potts looked at two particular regions in East Africa. One of them was Koobi Fora in Turkana in Kenya; the other, Olorgesailie Basin, also in Kenya.

Using data on climate variability collected by Peter de-Menocal, a geochemist based at Lamont-Doherty, Potts shows that the more adaptable *Homo rudolfensis* and *Homo ergaster* seemed to prosper during periods of great climate variability, while the more specialized *Australopithecus boesei* faltered. Fossil *Homo* remains first appear in Koobi Fora about 2 million years ago. This marks the beginning of a 250,000-year period of great climate instability following 400,000 years of relative stability. Potts notes that during this period roughly 75 percent of the fossil remains are *Homo* while 25 percent come from *boesei*. The more specialized *boesei* make a comeback during the following 60,000 years of climate stability, and then *Homo* fossils are resurgent when high climate variability returns between 1.65 million and 1.55 million years ago. *Boesei* recovers a bit during the next 30,000 years but then disappears.

Potts also looked at stone toolmaking down through time. It became a permanent feature of hominin groups starting about 1.85 million years ago, but Potts argues that toolmaking did not endure across environmental shifts until about 1 million years ago. This suggests that at some point toolmaking advanced to the point that it helped hominins survive extreme climate change. Climate challenges favored the generalists of successive hominin lines, and by about 1 million years ago those generalists had developed!enough tools to deal with the changing weather and survive the periods of instability. Potts bases his argument on the assumption that "adaptive versatility" is an evolvable trait.

Evidence in favor of this argument includes data suggesting that during periods when hominin species were proliferating, other plants and animals were changing rapidly as well. For instance, the fossil record of the Olorgesailie Basin shows that between 800,000 and 500,000 years ago, ancient forms of many big mammals disappeared, to be replaced by their modern descendants. Among others, he cites a zebra—*Equus oldowayensis;* an elephant—*Elephas recki;* a baboon—*Theropithecus oswaldi;* and a hippo—*Hippopotamus gorgops*—as creatures that became extinct during this period of climate upheaval. Potts notes that all of these animals were specialized grazers of the savannah who gave way to generalists, something that might be expected when the food supply is undergoing rapid change. He notes also that, at the same time, *homo* was going through a period of rapid brain growth, which was one evolutionary path toward becoming a better generalist. *Homo erectus* largely disappeared by the end of this period, after lasting more than 1 million years. The big jump in brain size between *erectus* and *Homo sapiens* took place in the neocortex, suggesting that by this time our ancestral line was fully committed to brainpower as a device for weathering climate changes.

By the time modern humans arrived, human brainpower had long since passed this threshold. This did not mean that early humans were immune to the effects of climate. For instance, archaeologists have found evidence in the Blombos Cave in coastal South Africa of relatively sophisticated bone awls, fishhooks, art, and organized fishing dating back 80,000 years or more. In 2004, a team of archaeologists led by Christopher Henshilwood of the University of Bergen in Norway found a cache of bored shell jewelry dated to 75,000 years ago*—30,000 years older than the date of beads made

Science, April 16, 2004.

from ostrich shells found in Kenya that, until this new find, were thought to be the first evidence of jewelry. The various findings from Blombos Cave bespeak a people who appeared to have developed complex social behavior tens of thousands of years earlier than had been thought.

The cultural development of these early Africans might have been interrupted by severe events such as the "volcanic winter" brought on by the eruption of Mount Toba roughly 71,000 years ago. That catastrophe—2,800 times more powerful than the explosion of Mount St. Helens—was the biggest eruption in the past 2 million years. Estimates are that the ash and gases thrown into the atmosphere lowered global temperatures between 3 and 5 degrees centigrade for the next six years and precipitated an ice age that lasted 1,000 years. It is easy to envision how an "instant Ice Age" might set back humanity's timetable for cultural development by tens of thousands of years. At this point in prehistory, the pace of advances in material culture was vastly slower than the explosion that took place with the advent of the first civilizations.

The notion that climate impacted human evolution and early material culture remains controversial. Ofer Bar-Yosef, the chairman of Harvard University's anthropology department and a proponent of the view that climate has played a key role in the rise and fall of civilizations, argues that climate has much more of an impact on complex societies than prehistoric cultural evolution. He argues that while different hunting-and-gathering groups may have been fatally impacted by climate change, other groups survived, and that it was only when societies became fixed in one place that they became more vulnerable to climate. He describes the Blombos setback as irrelevant to human cultural evolution because South Africa was a "cul-de-sac" far removed from the center of radiation of ideas and people in the Middle East and North Africa.

Speculating that a fixed civilization might be more vulnerable to climate shocks than those who can move with the game does not rule out the possibility that climate might set back human material evolution. One of Bar-Yosef's postulates holds that innovations start in one place and spread rapidly as different peoples adopt the practice. He says the archaeological record often seems to suggest the simultaneous discovery of an innovation across a wide area because the margin of error in dating techniques is larger than the time it takes for an idea to spread. To illustrate this, Bar-Yosef uses the example of McDonald's, the paradigm of fast food, which spread from a single shop to worldwide penetration in about fifty years. If some future archaeologist used carbon dating to analyze the shards of these fast-food restaurants, he or she could plausibly come to the conclusion that McDonald's sprang up everywhere around the world at once because C14 has at least an error of plus or minus fifty years. What's true for McDonald's might well be true for prehistory, and innovations lost or forgotten as early humans scattered to adapt to a changing climate might not be reinvented for many years.

If changing climate helped *Homo sapiens sapiens,* it was by wiping out close relatives that could not adapt as the weather and food supply shifted. Millions of hominins perished when climate turned harsh at different points.

3

Destroyer

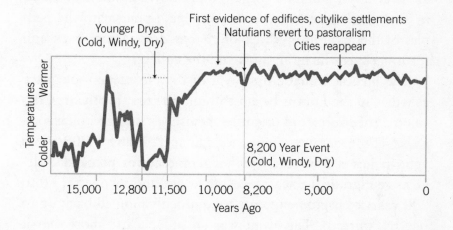

IT WOULD NOT BE until the Holocene period, which began roughly 10,000 years ago, that the survivors of the ice ages had a big enough brain, and an expanse of climate stability long enough, to nurture civilization and technology. The Eemian, a warm interval about 130,000 years ago, offered a period of climate stability when *Homo sapiens sapiens* was relatively young, but, at that point, material culture was vastly more rudimentary than when humans emerged from the glacial era into the Holocene and, as far as we know, it did not progress far enough during this earlier interregnum to leave any trace of organized settlements. Evidence from both the animal kingdom and human history contradicts the folk wisdom that "necessity is the mother of invention." Rather, it is during times of surplus that groups have the leisure and wherewithal to innovate (successfully adapting to climate challenges does not mean thriving during periods of rapid change). While we may be a product of the climate roller coaster, civilization only flourished during the brief pause at the top of the ride of the past 10,000 years. And, as we shall see, civilizations and empires have proved vulnerable even to the relatively minor perturbations of climate during the Holocene. Once humans began to put down permanent stakes in Turkey, climate began a game of ping-pong with the peoples of the area, as episodes of change drove tribes back and forth between the mountains and the valleys.

The biggest climatic upheaval of the Holocene took place as tribes in the Levant began to build the first protocities, develop agriculture, and organize themselves into complex societies. The event was an extraordinary cooling 8,200 years ago. In just a few years, global temperatures plunged 5 degrees centigrade, to nearly ice age levels, and then, after 60 to 200 years of frigid temperatures, suddenly jumped back up to relative warmth. The event was so sudden and short that it was over before glaciers had a chance to grow.

Just before this climate catastrophe, the world was a warm and pleasant place. Although this abrupt cooling took place a few thousand years after the beginning of the Holocene, there was still a good deal of ice remaining in the northern hemisphere in the form of remnants of the great ice sheets of the previous ice age. In North America, the combination of a retreating Laurentide Ice Sheet and steady warming created an enormous lake (named Lake Agassiz after the Swiss scientist who first postulated that ice sheets once covered Europe) dammed behind the ice wall. As the warming continued, the lake became gigantic—perhaps 700 miles long and 200 miles wide—double the size of the Caspian Sea and dwarfing the size of all the present-day great lakes combined. Eventually, because of warming or some movement in the ice sheet, a hole opened in this ice dam, and the titanic pressures of the vast lake unleashed what was probably the greatest flood of the past 10,000 years. Much of the lake, which had begun forming 4,000 years earlier, drained in a matter of months.

With the force of 80 to 100 Amazons (between five and ten million cubic meters per second), water poured out, probably into Hudson Bay and eventually into the North Atlantic. Writing in *Science* in 2003, Garry Clarke, David Leverington, and colleagues speculated that as the climate warmed and the lake grew, the water eventually tunneled through the fast-disintegrating ice sheet. The idea is that this sudden slug of fresh water changed the salinity, lightening the surface water to the point that it no longer sank in the Labrador and Norwegian seas. While the cold event of 8200 B.P. shows up in virtually every record as the largest climate disruption of the Holocene, the evidence is not yet conclusive as to whether the flood caused a shutdown of the thermohaline circulation. Clarke and his colleagues note that climate models support this hypothesis, but the physical evidence of proxies remains ambiguous. Some critics note that while they haven't found evidence of this massive flood exiting Hudson Bay, it's not out of the question that the flood took a different exit to the oceans, such as the St. Lawrence Seaway.

Less ambiguous is evidence of sudden cooling around the world. Richard Alley says that much of the world was a colder, drier, and windier place as the global climate adjusted to the shocks of this superflood. Evidence suggests that it was windy in much of the U.S. South, as well as in parts of South America. It was cold in West Africa, and very cold and windy in the cradle of civilization, particularly Turkey, where agriculture and the first large villages were just being established. The 8200 B.P. cold event stopped this process in its tracks, interrupting the progress of the first stirrings of civilization.

This is the argument advanced by Bar-Yosef. While he may be dubious about the relevance of Blombos Cave to human material evolution, he is a powerful advocate of the role of climate in the fortunes of human societies. Much of his work has involved excavations of Natufian sites in the Levant, the stretch of the Middle East that runs some 1,100 kilometers from the south side of the Taurus Mountains in Turkey to the Sinai Peninsula. The Natufians were a Neolithic culture that extended from modern-day Israel to southeastern Turkey, and around 11,500 years ago they were among the first peoples to start cultivating grains and domesticating cattle, sheep, goats, and pigs in this landscape, which varied from Mediterranean woodland to scrubland and steppe. They developed this expertise either at the end, or just following the end, of the Younger Dryas cold period. By 9500 B.P., they had settled in large villages and erected stone structures up to 10 feet tall. These early edifices were not the pyramids or gates of Troy, but still represented a major step from the animal-skin tents of most other peoples on the planet.

Then came the superflood on a faraway continent, and the sudden, howling, desiccating winds of 8200 B.P., after which, says Bar-Yosef, everything came crashing down. Crops died, growing seasons shrank, and villages emptied as Natufians turned to pastoralism. Most did not revert to hunting and gathering, he says, but set off with their newly domesticated animals in search of forage and food. It was not for another 3,000 years that the residents of the Levant made a fresh start at developing complex societies with their monuments and cities.

Bar-Yosef argues that these groups reverted to pastoralism because early complex societies had no buffer against a shock like climate change. Indeed, the anthropologist asserts that only complex societies are truly vulnerable to climate change because they are fixed in place. Hunters and gatherers suffer, but they can follow the food supply. As Bar-Yosef said during an interview, "Without a FEMA [Federal Emergency Management Agency] or someone to say 'Here's some grain, stay where you are,' these early agricultural societies could not absorb a major climate shock."

The sudden cooling of 8,200 years ago may have interrupted the development of cities, but it may have been the next big cooling that restarted the process. As Bar-Yosef interprets the record, some of the tribes who abandoned the Fertile Crescent eventually followed the game into the mountains and they resumed their nomadic ways. Three thousand years later, another sudden cooling may have driven them back down from the mountains into the flat lands where they built the first Mesopotamian civilization. This hypothesis derives in part from a chance meeting between a glaciologist and a Yale archaeologist.

Lonnie Thompson, the Ohio State–based glaciologist, has reputedly spent more time above 18,000 feet in altitude than anyone on the planet. It's at those heights in the Himalayas, on Mount Kilimanjaro, and in the Andes that he has found ancient glaciers with high-resolution records of climate in the tropics and mid-latitudes. Throughout this period, he's watched in dismay as these glaciers have retreated with alarming speed as the climate has warmed. In recent years, the speed of melting in some of the ice caps and glaciers has doubled. On Kilimanjaro, Thompson has uncovered bubble structures in the ice cap suggestive of melting and refreezing in recent years, something that doesn't occur at any other time in the entire 11,700 record.

(Typically, glaciers at that altitude lose ice through a process called sublimation in which sunlight transforms ice into water vapor without melting.)

Thompson quotes the Antarctic explorer Ernest Shackleton's comment "What the ice gets, the ice keeps." Lately, however, because of melting, the ice has been giving back some of the objects it has been hoarding for thousands of years. For instance, in 2002 the receding ice of the Quelccaya Ice Cap in the Andes uncovered the frozen remains of a plant called *Distichia muscoides*. What struck Thompson was that the plant showed virtually no signs of decay, but that was just the beginning of the story.

Botanists recognized the flower as typical of those that currently live at altitudes nearly 1,000 feet lower than where it was discovered, suggesting that just before it was covered by ice, the climate had been quite warm. Then Thompson sent the plant to the Lawrence Livermore National Laboratory for radiocarbon dating and DNA analysis. Before he got his results, he got a bemused call from the scientist doing the work. He told Thompson that the lab treats organic material before doing its analysis, and that when a plant dissolves during the treatment, the rule of thumb is that the material is too young to date. That's what happened to the plant that Thompson sent to the lab, yet the tests showed it to be 5,177 years old (other plants buried earlier by the ice dated as far back as 50,000 years).

It was remarkable in itself that a soft-bodied plant remained preserved without any evidence of decay for more than 5,000 years, but what its preservation and the call from Lawrence Livermore suggested to Thompson was that the climate shifted from sunny to frigid so rapidly in the Andes that the plant was frozen virtually overnight. It remained frozen and covered with ice for the next 5,200 years.

Now let's shift several thousand miles to the Alps near the

Austrian and Italian borders in 1991. There at the edge of a retreating glacier, a group of hikers found the preserved remains of a man resting against a boulder with an arrow wound in his back. Again, the remains were so well preserved that the Austrian authorities initially treated it as a murder investigation. The ancient clothing ultimately convinced the government to call in archaeologists. Carbon dating placed the time of death about 5,175 years ago (with a margin of error of plus or minus 135 years). What intrigued Thompson was not the foul play but that, almost as soon as the man died, his body was frozen in place and remained buried by ice and preserved for the next 5,200 years.

The two pieces of evidence, confirmed by numerous proxies taken from glaciers, caves, and tree rings around the world, attest to dramatic climate change around the world at that time. Lakes in Africa dried up (including one in the Sahara that has never subsequently refilled). Tree rings from England dated to 5194 B.P. show that year to be one of the warmest in the tree ring record. Then, just a few years later, methane levels recovered from ice cores in Antarctica from a bit later show that biological activity around the world dropped precipitously. It's as though a switch was flicked.

Because of his pioneering work at extremely high altitude, Thompson has become a celebrity in academic circles and does a good deal of speaking at universities and conferences. Some years back, he gave a talk at Yale in the course of which he covered the very abrupt and extreme cooling of 5,200 years ago, and after mentioning the flower and the iceman, he wondered aloud what else was going on at that time. He recalls that a hand shot up in the audience and an archaeologist named Harvey Weiss stood up. Weiss pointed out that it was about then that the first cities emerged in Mesopotamia as people moved down from the mountains.

The events of 5,200 years ago left a profound imprint on

people around the world. Thompson notes that Thor Heyerdahl noticed that it was about this time that calendars were invented on four different continents. Could it be that the shock of massive upheaval in weather spurred far-flung peoples to realize that they could not take the clockwork regularity of the seasons for granted?

Archaeologists had long speculated on what had forced these formerly dispersed bands of people to join together in cities in the lowlands. Thompson may have supplied one obvious candidate: an abrupt and protracted descent into much colder temperatures (I'll get into the possible explanations for the event in a later chapter). Thompson also supplied the beginning of a story to which Harvey Weiss already had the end. At the time of Thompson's talk, Weiss had already spent more than a decade trying to prove that another episode of cooling and drying 4,200 years ago had led to the demise of the Akkadian civilization that began when tribes had moved down from the mountains 1,000 years earlier. For Mesopotamia, this was to be the last episode of climate ping-pong between the mountains and the valleys.

4

The First Victim

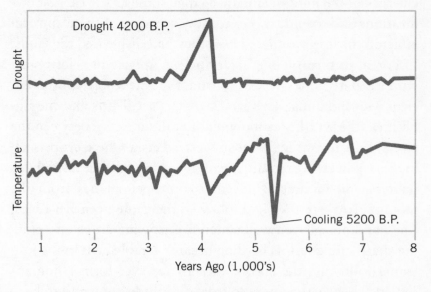

SIGNALS OF DROUGHT FROM MT. KILIMANJARO

THE CLIMATE EVENT that may have killed off the Akkadian civilization 4,200 years ago, was far less severe than the earlier sudden global cooling that helped create it. Still, it was vastly more severe than anything experienced in recent times. By way of comparison, consider the typical elements of a strong El Niño: drought in the Middle East, India, China, Australia, Indonesia, and parts of Brazil; floods in California and parts of China; and generally unsettled weather around the world. In

1997–98, a strong El Niño inflicted about $100 billion in damage. Now imagine if an El Niño continued month after month, year after year for a century, two centuries, and then decades more. Imagine further that the weather disruptions were many times more extreme, with incessant howling winds scouring topsoil from fertile lands and driving people to madness. Imagine such a relentless series of windstorms and droughts afflicting generation after generation after generation, and a picture emerges of the kind of climate change some of the earliest civilizations had to endure. No wonder so many of these climate-afflicted cultures wondered how they had displeased the Gods.

For a picture of one such ancient apocalypse, let's start with the bare bones of a story from the Fertile Crescent in ancient Mesopotamia. This land between the Tigris and the Euphrates (the word Mesopotamia literally means "between the rivers") is one of the places in western Asia where crops such as barley were first cultivated in an organized fashion as hunters and gatherers moved to the open planes from the mountains where they had followed the game in earlier times. The flat lands proved well suited for growing crops and grazing sheep and goats as herding became established. In the millennia following the end of the last ice age, former nomads settled. They learned how to store and trade grain and to keep accounts of who sold what to whom; they learned how to combine tin with copper and make bronze tools and weapons. Facilitating these first experiments in what we call early Bronze Age civilization in the third millennium B.C. was a relatively benign climate as westerly winds brought moisture from upwelling (the rise of cool deep water) in the eastern Mediterranean. In the northern part of this region, which included parts of present-day Syria and Iraq, rainfall was sufficient to support dry-land farming, while farther south, farming societies figured out how to divert the Euphrates to irrigate the drier, hotter lands.

The tribes that moved down to the plains retained their truculent attitudes and the warlords saw new opportunities for power and plunder in the settlements as they became organized and prospered. By 2340 B.C., an enterprising warrior king from southern Mesopotamia named Sargon had conquered a sufficient number of the independent city-states across a vast swath of the Middle East to legitimately lay claim to one of the planet's first empires. Based in the lost city of Akkad, the regional power came to be called the Akkadian Empire. Its origins are obscure; its language is the earliest Semitic language yet identified. Akkad was not, however, primitive: the empire's written language and well-developed system of accounts testify that the people had a sophisticated social organization when they first invaded the Habur Plain.

At its peak, Akkad ruled such fabled places as Nineveh, Ebla, and Ur. As Harvey Weiss described it in *The Sciences,* "The Akkadian Empire controlled trade from the silver mines of Anatolia to the lapis lazuli mines of Badakhshan, from the cedar forests of Lebanon to the Gulf of Oman. In northern Mesopotamia, meanwhile, fortresses were built to control imperial wheat production. To the south, irrigation canals were extended, a new bureaucracy established and palaces and temples built from imperial taxes."

Laborers were paid in wages of two flat-bottomed *sila* bowls a day of lentils and barley. (The bowls held almost exactly a liter, says Weiss, a unit of measure that was not formalized until the French Revolution, 4,000 years later.) They built storehouses, schools, and fortifications to defend imperial cities.

One of the most prosperous of these outposts was a place named Tell Leilan by modern archaeologists (a tell is a mound containing the buried detritus of ancient cities), located on the Habur (sometimes spelled Khabur) Plain between the storied Tigris and Euphrates rivers in northeastern Syria, and not far from the border of present-day Iraq. By 2280 B.C. when the Akkadians took over, Tell Leilan had already been a city for more than three hundred years. By 2200 B.C., Tell Leilan had been an urban area roughly as long as New York City had in 2006.

Priests ruled the roost in these early cities, until warrior kings interposed themselves as the interpreters of the gods' wishes. As the Akkadians consolidated control of the area, they began an ambitious urban renewal program, erasing many of the structures of the previous government, and erecting new cultic platforms for animal sacrifices. They built moats for defense, storage rooms for grain, and administrative buildings for bureaucrats. They built a one-room schoolhouse where little Akkadians could learn their glyphs and measures (they learned a base 60 number system that we still use today in the measurement of time and angles). A basic building material was basalt, which laborers hauled from an outcrop 20 kilometers away. Some of these blocks contained 4 to 5 cubic meters of rock.

As Weiss reconstructed the story from the evidence of his dig, sometime just before 2200 B.C. work began on a new building inside the main acropolis, situated hard by a small house. Laborers would haul the blocks up to the top of the acropolis, where skilled workers would dress the blocks to be fitted to the extended walls of the complex. Once the basalt was fitted, other workers would top the basalt with mud, potsherds, and bricks.

One day, when workers were constructing a wall in part of the acropolis, work simply stopped. Half-dressed blocks were abandoned; other materials were left scattered about. Workmen had dropped what they were working on and left in a hurry.

What were they fleeing? Invaders? A plague? Most probably they were fleeing starvation. Those left behind did not fare

very well. Excavations reveal a very large spike in infant buri-
als in the years after 2200 B.C. Building that wall was possibly
the last act of construction in Tell Leilan for the next three
hundred years.

The diaspora marked by the abandonment of the building
site was one of the last acts of a drama that had probably
begun unfolding several years earlier. The first sign of the
coming bad times was the failure of the moist summer winds
that had first lured settlers to the Habur Plain centuries earlier.
Instead of rain, the Akkadians and their subjects were baked
by dry, hot winds from the north. Precipitation fell by 30 percent, and crops withered in the field; the raw wind picked up
the topsoil and blew it south. With their surplus dwindling,
the fields barren, and laborers consuming what was left, at
some point the Akkadians decided the game was up. The
bleak despair of the time was captured in a lamentation entitled "The Curse of Akkad" that was probably first written on
tablets a hundred years later:

> The large fields and acres produced no grain
> The flooded fields produced no fish
> The watered gardens produced no honey and wine
> The heavy clouds did not rain
> On its plains where grew fine plants
> "lamentation reeds" now grow.

Many of the Akkadians moved south, likely as word filtered
back that the Euphrates, though diminished, continued to
flow, supplying irrigation water to the fields there (the northern part of the empire, previously watered by adequate rainfall, had no irrigation infrastructure). Some refugees became
pastoral nomads, moving with their herds in search of fodder.
In the years following the collapse of Tell Leilan, references to
"barbarians" from the north and to other strange names

began appearing in the documents of the UR-III Empire that assumed control of the southern part of the former Akkadian lands.

Those who remained at Tell Leilan left no trace. As the decades went by, sand and dust gradually entombed the acropolis. When the winds and drought finally abated, some three hundred years later, new settlers moved in, but who knows what traces remained of humanity's first great empire when, in about 1800 B.C., Shamshi-Adad I, the Assyrian king, decided to establish his capital, Shubat Enlil, on what is now called Tell Leilan.

As the tragedy of Tell Leilan unfolded, others were suffering too. In the Indus Valley, two cities, Mohenjo-daro and Harappa, spiraled into catastrophic decline between 2200 and 2100 B.C. In Egypt, the Nile shriveled to the point that scribes report people walking across the riverbed. The drought punctuated the end of the Old Kingdom. Another contemporaneous collapse of an early Bronze Age civilization unfolded in Crete. The droughts extended far beyond the cradle of civilization, parching Africa and desiccating some of the wettest places on the planet, such as the Amazon, which suffered the worst drought of the previous 17,000 years, according to recent reconstructions of its climate done by Lonnie Thompson.

We may never know what miseries this event brought to the indigenous tribes of South America or Africa because they left no retrievable records or accounts. The images in the accounts of those who did leave records evoke a dark age that interrupted the first stirrings of civilization, loosing tides of refugees across great swaths of the globe and, along with these refugees, the bandits and warlords who fed on chaos.

When archaeologists first began finding evidence of these synchronous collapses, decades ago, the hand of climate was well concealed, and most ancient histories looked for more prosaic culprits. Now Weiss, Thompson, and others have gathered evidence that suggests this ancient apocalypse, like the climate upheaval that first caused the tribes to move into the Habur Plain, represents the first act of a serial killer that has stalked civilizations ever since.

5

Weapons of Mass Destruction: Disease, Migration, Conflict, and Famine

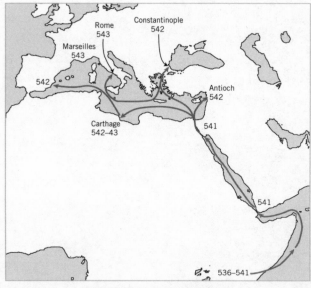

FIRST WAVE OF JUSTINIAN PLAGUE

As WE MOVE closer to the present, historians can draw on contemporary accounts, tax rolls, and other detailed records to fill out the picture presented by proxy records. These records allow us to see the ways in which climate has both direct and indirect effects on the fortunes of a society. A good example of the ways in which sudden climate change unleashed a cascade of repercussions took place in A.D. 536, when some mysterious event darkened the sun for over a year. It's unclear what caused this event, but the immediate global impacts included a drought in Mesoamerica that foreshadowed the end of the Mayan civilization four hundred years later, and famines in Asia. Indirect events included waves of migration and disease, perhaps including the so-called Justinian Plague, a late-Roman-era pandemic whose successive outbreaks killed millions during the following two centuries. The A.D. 536 event had such profound consequences that some historians argue that it marks the dividing line between the classical and modern eras.

Joel Gunn, an anthropologist at the University of North Carolina, has explored its global dimensions. It shows up in tree rings, ice proxies, and other mute records, as well as in letters, accounts, and folklore from the British Isles to the Roman Empire to China and Japan. The tree ring and ice core data point to a sudden cooling in England, Sweden, Finland, Siberia, China, and Korea, among other places, and drought in Peru, East Africa, India, China, and Korea.

Gunn unearthed a detailed account of what occurred in 536 in the letter of a praetorian prefect in Italy to his subordinate. The prefect, Magnus Aurelius Cassiodorus Senator, wrote:

Just as there is a certain security in noting seasons recurring at their proper times, so likewise we are filled with great curiosity when such events seem to be altered. What sort of experience is it, I ask you, to look upon the principal star and not to perceive its usual light? To look upon the moon—the decoration of the night—in all its fullness but without its natural splendor? We all perceive a blue colored sun. We wonder that at noon bodies do not have shadows, that the strongest heat has reached the inertia of extreme tepidity, because—not by the momentary failure of an eclipse, but for the space of an entire year—it has failed to be fixed in its course. . . . Thus we had winter without storms, spring without moderate temperature, summer without heat. How can we hope for a temperate climate when the months which could have ripened the fruits froze them instead by its northern blasts? How can the earth provide fertility if it is not warmed by the summer months? How can the grain sprout if the soil has had no rain?"

In fact, the answer to Cassiodorus's question is that grain couldn't sprout, crops were ruined, and people died by the millions.

According to David Keys, author of *Catastrophe: An Investigation into the Origins of the Modern World,* the cause of this particular crisis is revealed in other accounts from the era.* A Roman official wrote of a "dry fog" and frigid temperatures. A Roman historian wrote of a "dim sun" that remained darkened for eighteen months. From the British Isles came accounts of "colored rain"; from a Chinese history there are accounts of "yellow dust that rained down like snow" and could be picked up by the handful.

With the possible exception of the yellow dust, the accounts suggest that some huge explosion poured enormous amounts of material into the air, creating a "nuclear winter," as was described for Mount Toba, which erupted tens of thousands of years earlier. The absence of any clear record of such an eruption has led some to suggest the impact of a comet, asteroid, or the airburst of some large extraterrestrial object. Keys believes that he found the culprit in two calderas just under the sea surface between Java and Sumatra in Indonesia

*David Keys. *Catastrophe: An Investigation into the Origins of the Modern World* (Arrow Books, 2000).

not far from Krakatoa, which memorably blew itself apart on August 26, 1883. He cites an account of an island being split in half cited in the Javanese chronicle *The Book of the Ancient Kings*. The dates don't match up, but Keys argues that the discrepancy might be the product of rewritings and retellings, as political upheaval created an historical version of the telephone game down through the ages. Whatever the cause, it seems likely that this particular cooling was the product of an extraordinary event and not a climate cycle.

Keys does not shy away from big ideas. In *Catastrophe*, he argues that by unleashing the plague from the south and causing barbarians to move westward from Asia, the climate upheavals of 536 played a key role in the end of the Roman Empire, the rise of Islam, and other events that marked the end of ancient times and set the stage for the emergence of the modern world. It's a sweeping claim, so bold that it almost begs contradiction, even from those willing to posit the consequential role of climate in human affairs. When a civilization is in decline, after all, the agent of its end might be entirely different from the cause of its decline. When pneumonia kills an aged patient in failing health, is pneumonia the killer, or merely the final nudge? As Keys exhaustively documents, by the sixth century, the Roman Empire was senescent, with little sense of mission or purpose, kept together principally by the fear of the gathering barbarians at the gates who were constantly probing for opportunities and signs of weakness.

Keys is quite persuasive in his arguments about the forces that put the barbarians at the gates in the first place. Thousands of miles east of the Roman Empire, on the steppes of Mongolia, the drastic climate upheavals of the years following 536 shifted the balance of power from a warrior culture based on horses to their former vassals who were pastoralists raising cattle. The horsemen were Avars, while the vassals were indigenous Turkic people. As Keys tells it, the Avars had ruled the roost for 150 years until about A.D. 545, when their grip on power began to fail. The reason, according to Keys, is that severe drought favored the cattlemen over the horsemen. Cattle, with a rumen positioned so it that can ferment and liberate the protein in the plant's cell walls before it passes into the small intestine, can make far more efficient use of scarce fodder in times of dearth. As their horses died in the years after 536, the Avars were successively stripped of their weapons, their meat source, and their principal currency for barter.

This is how Keys describes the consequences: "First—in A.D. 545—the Turks snubbed their official overlords, the now much-weakened Avars, by establishing direct diplomatic links with the imperial government of northern China. Then, in 551, the Turks virtually saved the by now feeble Avars from destruction by rebel tribes. And in the following year, the Turks pushed for political equality with their Avar masters by demanding that the Avar ruler give the Turkic kagan (king) one of his daughters as a wife. The proud Avar leader refused and the Turks used his refusal as a pretext for overthrowing Avar rule."

After a bloodbath in which the Avar leader, Anagui, committed suicide to avert a humiliating ritual execution, thousands of Avars decided to take to the road to avoid enslavement or worse. They headed west with their remaining horses in ox-drawn caravans, arriving at the edges of the Roman Empire around A.D. 557, roughly five years later. The trip could have taken less time, but the lands between Mongolia and Eastern Europe were occupied, and the Avars had to fight their way through parts of the territory. Apparently, although the Avars were shaken by their defeat, they were also stirred. Conscripting vassal peoples such as the Slavs along the way, they had regained their confidence by the time they reached the outer edges of Hungary and were up to their old tricks, invading lands and extorting protection money. Keys calculates that over fifty years the Avars extracted about $7 billion in today's money from the weakened Roman rulers.

Keys makes a good case for the role of climate in this migration (nearly a hundred years earlier, the controversial Yale historian Ellsworth Huntington made a similar case in *The Pulse of Asia**). Would the Avars have lost their grip and headed west without the climate surprise of A.D. 536? Certainly, they would have eventually lost control of Mongolia. Even if the Avars had not shown up in Hungary, however, there were plenty of other candidates who would have eventually become sufficiently emboldened to challenge the empire.

At the same time, the empire was being weakened from within by reverberations of the 536 events, one of which was the unleashing of the Justinian Plague. Changing climate can unleash disease in a variety of ways. When precipitation patterns and temperatures change, for instance, creatures in the affected ecosystem must adapt. Typically, the organisms that adapt the fastest are those that have short generations and many offspring—so called R-strategists. These include microbes, pests such as cockroaches or rats, and other "weedy" species. It's a blind numbers game: if change increases the food supply, populations can expand rapidly. If, on the other hand, change diminishes or changes food availability, the R-strategists have masses of offspring and short generations, so a successful adaptation can rapidly take hold.

The animals that prey on these weedy species typically have fewer offspring and longer generations since they have fewer enemies in normal times. This makes them either less able or slower to adapt when climate pulls out the rug. During and after rapid changes of climate, it's possible to have a temporary situation in which pest numbers explode as either food increases, the pests adapt, predators temporarily diminish, or

*Ellsworth Huntington. *The Pulse of Asia: A Journey in Central Asia Illustrating the Geographic Basis of History* (Houghton Mifflin, 1919).

some combination of these elements. This is the argument advanced by a number of epidemiologists, notably Paul Epstein, associate director of Harvard University's Center for Health and the Global Environment.

With climate changing rapidly today, Epstein and others have been able to document a number of surprising ways in which this interplay between climate change and disease plays out. Epstein points out that among the biggest beneficiaries of recent warming have been bark beetles, which are sweeping up from New Mexico to Alaska. In the lower states, the beetle numbers have surged because warmer temperatures are allowing them to overwinter so that they produce two generations instead of one. In Alaska, where colder temperatures formerly kept them at bay, they are flourishing. In Montana, warming temperatures have allowed them to invade new species of trees at higher altitudes. They have broadened their preferred diet from lodge pole at lower altitude to include white pine at 8,000 feet and above.

As warming provides a cozier environment for the beetles, accompanying drought also makes the trees more vulnerable by drying out the rosin or pitch. After about a year, the combination of stresses kills the trees, which then become more fuel for fires. This is one reason why the American West has witnessed a steep increase in the number of wild fires.

What works for bark beetles also works for the microbes that affect people. Warming allows mosquitoes to move up to higher altitudes where diseases such as dengue fever and malaria can invade new populations. One well-studied example of the interplay of microbes and climate comes from the El Niño of 1993, which played a role in an outbreak of hantavirus in the American Southwest. The scenario, as described by Jonathan Patz of Johns Hopkins University in Baltimore, went as follows: First El Niño–related floods brought water to desert regions. This dramatically increased the food supply for deer mice, whose numbers exploded by more than an order of magnitude. The hantavirus likes riding around in deer mice, and the virus can get from mouse to human through mouse excrement, which dries and then gets carried into human lungs as dust. As great numbers of deer mice found shelter in homes, the opportunity for the virus to jump to people increased dramatically.

According to Keys and others, something similar happened in the years following the big chill of A.D. 536 in Africa. In this case, the disease was the plague, and the reservoirs for the disease were probably two rodents from East Africa—gerbils and multimammate mice. Both exemplify the definition of R-strategists. The gerbil can produce two litters of ten offspring a year, while during times of abundance the multimammate mouse can compound its numbers by tripling its litter size from five to fifteen. Citing arguments produced by the Centers for Disease Control and Prevention, Keys argues that either flood followed by drought, or drought followed by flood, can tilt the predator-prey balance in favor of the R-strategists over their predators, which tend to be K-strategists—creatures that have fewer offspring and invest more in their survival. It is highly likely that the events of 536 produced some version of this scenario.

The animals themselves are immune to plague, but carry the fleas that carry the disease. The plague makes the fleas sick by blocking their digestive tract and they respond by becoming ravenous, biting everything in sight. If developed by humans, the strategy would seem fiendishly ingenious, but the microbes developed it through simple trial and error.

Neither the gerbils nor the multimammate mice ordinarily have much contact with humans, but when conditions are right, they might have contact with another rodent, notes Keys, called *Arvicanthus*. This rat, not immune to plague, is right up there with the gerbils as a champion breeder, and it could serve as a bridge for the plague because it does have contact with *Rattus rattus,* the black rat, which for millennia has made a good living off of the leavings of human settlements, and, equally important, ships. Once the fleas made the jump to black rats, humans, through trade and empire, opened vast new worlds for the microbe to conquer.

The free-riding plague microbes seized the day. They traveled up the coast of Africa with the ivory trade and radiated outward from East Africa wherever rats flourished. After laying waste to Constantinople and Alexandria, among many other places, the plague made it to Great Britain around 549. Its decimating impact on the aristocracy shows up in several records, and Keys argues that it led to the virtual abandonment of Tintagel, revered because legend has it that it was in Tintagel that King Arthur was conceived. Arthur himself likely died during the upheavals of climate and disease in the years that followed 536.

The disease also fanned out across Africa. While the plague's impact on the Middle East and Europe has been well documented, what happened in Africa, particularly the Congo Basin, remains an intriguing mystery. During the past few years, botanists, archaeologists, and explorers have been finding evidence of a massive collapse of human numbers in sub-Saharan Africa right around the time that the A.D. 536 cooling wreaked havoc in the rest of the world. In this case, the evidence is as much ecological as archaeological.

I first heard of this collapse in the early 1990s. In 1992, I'd ventured into the Ndoki rainforest, an extremely remote forest in the northern part of the Democratic Republic of the Congo. Girded by swamps and wild rivers, it was so inaccessible that even the Pygmies did not know about it. The animals were naïve of humans, and some of the shyest and most hunted creatures in Africa, such as forest pigs and yellow-backed duikers, would simply stare at us when we walked by.

At the time, Mike Fay, the botanist-explorer I accompanied, and the few other scientists who probed this forest had found no evidence that humans have ever entered the Ndoki.

A couple of years after this trip, Fay contacted me with news. While exploring riverbeds and pits dug by elephants, he had found ancient palm nuts. The nuts are evidence that humans had been in the area in the past. When he finally dated the nuts, he discovered that people had lived in the Ndoki over a thousand years before we entered the forest. Their numbers had peaked about A.D. 400 and declined thereafter. By the turn of the first millennium, no one remained.

Fay wrote about this finding in his dissertation, in which he also offered a hypothesis for the subsequent collapse. The evidence suggests that Bantu peoples first started migrating from grasslands in Cameroon into Central Africa about 5,000 years ago, the earliest populations being Stone Age hunter-gatherers. Iron use and agriculture worked its way into the area either through adoption or conquest, particularly in the years beginning about 1900 B.P. Long before this, people had begun cultivating oil palm (*Elaeis guineensis*), which, Fay argues, occurs in Central Africa only because people brought it there. Without humans clearing areas and preparing fields, the palms can't survive in the jungle, which makes them an ideal marker of past civilizations since the number of nuts correlates well to human populations, and the palms themselves tend to disappear not long after the people. Along streams, however, there may be sufficient open area for the palm to persist without people for a while, and to be spread by animals; factors that could make it appear that humans persisted longer in an area than was the case.

Climate change may have abetted Bantu entrance into the Ndoki as well as hustled them off the stage. Periods of drying in prehistory could have opened the forest enough to permit the agricultural Bantu peoples to establish themselves. The only native African peoples who flourish in undisturbed forest are the Pygmies and other hunter-gatherers, and even these supremely adapted cultures never grow to large densities since they are limited by the availability of game in a particular area.

Once established in a degraded forest, the Bantu settlements energetically set to work pushing the forest back. Fay hypothesizes that a Bantu population explosion deforested an enormous swath of Central African forest stretching from Gabon to much of the Congo, and that, while the forest subsequently regenerated after the mysterious collapse of this population, evidence of the earlier damage persists in the relatively low diversity of the forest today.

What caused the collapse? Looking at the ecological evidence, Fay tends to think that the Bantu population outran the carrying capacity of the region. Much of the region has very poor soil, and without a constant supply of decaying forest vegetation, fields quickly become hard and barren. As food supplies diminished, people succumbed to war and famine as chieftanships fought to survive. It's a theory into which a little shove by a cooling and drying after A.D. 536 fits nicely. The Ndoki lies on the northern edge of the range of wet tropical forest in Central Africa, and any southward shift of the Intertropical Convergence Zone would dry the forest out in a hurry. Studies by Daniel Nepstad of Woods Hole Research Center and others have shown that tropical forests depend largely on rainfall and desiccate rapidly during times of drought. Ironically, this can leave wet tropical forests more vulnerable to fire than northern evergreen forests, where fire is a regular part of the ecology of the system.

Lee White, who uncovered palm nut evidence of the ancient rise and fall of Bantu populations in Gabon and Nigeria, wonders whether the collapse might have been brought about by disease. His favored theory is that climate change and agriculture opened a corridor allowing sleeping sickness and river blindness to travel into the central African settlements. And then, of course, there is the possibility that the plague found its way from East Africa into the heart of the rainforest.

If so, it would likely have come about through a chain of contact between farming peoples. Keys makes an interesting point when he argues that nomadic people largely escaped the plague because they offered less opportunity for rats to get established. Sedentary farmers were more exposed. The rainforest tends to efficiently digest human remains (which is why, as an aside, paleontologists have such a hard time reconstructing the ancestral forms of forest apes such as chimp, bonobo, and gorilla), so this argument is likely to remain speculative.

What's clear though is that when climate changes the context for people and ecosystems, both react, and the derivative effects of these interactions become complicated and unpredictable. While the impact of the climate disruptions in the years following 536 quieted within a decade or so, pulses of plague continued for the next two hundred years. Similarly, the droughts that helped put the Avars on the move also subsided after a while, but, if Keys and others are correct, the human migrations they unleashed played a crucial role in the final collapse of the ancient order. The millions who died in Europe, Asia, and the Congo from plague and famine also underscore a human vulnerability that civilization and all its safety nets have not been able to erase. For all our adaptability, we tend to crash with the other K-strategists when climate throws a curve ball.

6

Empty Promises of Water: The Collapse of the Mayans

Scale
One centimeter (1 cm) = 5 years

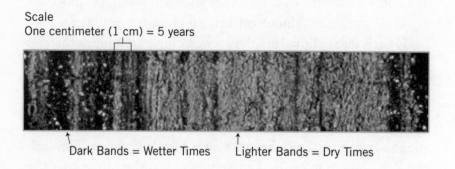

Dark Bands = Wetter Times   Lighter Bands = Dry Times

LAKE BOTTOM SEDIMENT CORE FROM YUCATAN PENINSULA

CHANGING CLIMATE can be violent and catastrophic, but even slower and more subtle change can decisively impact the structure of a society, undermining its ability to adapt. If a ruling class bases its legitimacy and privileges on its ties to the gods that provide water, then when the water dries up, the authority of the ruling class evaporates as well. The subsequent resulting chaos then amplifies famine, disease, and civil conflict. The historian looking only at data on climate and disease would see a correlation between climate change and a civilization's end, but a key element in the decline has been the collapse of government authority. With the collapse of authority, markets break down as well, and a society loses the resilience that might buffer people against the effects of a drought or epidemic. This might explain the collapse of the Mayan civilization 1,100 years ago. This argument faces a strong headwind, however, because it seems that almost everyone has a different theory for the decline of the Mayans.

Richardson B. Gill, an independent archaeologist uncovered over a hundred different theories of the Mayan collapse in the course of researching his book, *The Great Maya Droughts: Water, Life, and Death.* * Most of the theories, he wrote, had to do with the internal dynamics of the civilization itself, and almost all the theories involved catastrophic effects of human actions. "In short," as he puts it, the lesson of the Mayans seems to be that "the Collapse was their own fault."

Theorizing that the Mayans were done in by their own failings is especially tempting given that the society gloried in human sacrifice and ritual torture. Putting aside disapproval of its brutal rituals, however, the Mayan civilization would rank high as a success for its technological advances and its durability. Though they never exploited the wheel or metal tools, the Mayans built elaborate pyramids and structures, developed a sophisticated calendar and astronomy, and were masters at water management. They maintained their traditions, social organization, and power for at least 1,200 years despite an environment that was short on topsoil and water and lay dead in the path of Caribbean hurricanes.

The Mayans did many things right. Still, much of their history remains alien, even counterintuitive to a western mind. Notwithstanding their failure to discover the use of metals or beasts of burden, they built their largest cities in areas far removed from strategic resources such as water. But when the biggest of the southern Mayan centers of power collapsed in

*Richardson B. Gill. *The Great Maya Droughts: Water, Life, and Death* (University of New Mexico Press, 2000).

the years between about A.D. 850 and A.D. 950, they were gone for good.

Since the dawn of advanced civilizations, dynasties have fallen, notes political theorist Lisa Lucero of New Mexico State University, but usually a new dynasty emerges to rebuild. Not so in the case of the Mayans. The last stele was completed in the Mayan center of power Tikal in A.D. 909, and there is no evidence that anything more was ever built at the site, which had been previously occupied for over 1,000 years. Coincident with the fall of the dynasty, the city was largely abandoned. Why? Lucero believes that water is a common thread that explains both the siting of the big southern centers of Mayan power and their subsequent collapse.

This explanation has been offered before in the long list of reasons for the decline of the Mayans, a list that includes class strife, overpopulation, disease (yellow fever), deforestation and erosion, economic mismanagement, and a breakdown of trading relationships. Here is a case, however, where new evidence, specifically an increasingly detailed sediment proxy record may tip the debate. New seabed sediment proxy records, along with numerous supporting proxies, point to a series of extreme droughts in the years during which classic Mayan civilization collapsed. As Gill puts it in *Great Maya Droughts,* "the basic, fundamental premise of this book is that when a society runs out of food and water, the people die."

Water loomed over every aspect of Mayan civilization. At its peak, its centers of power and the authority of its rulers were completely leveraged on its continued availability. But unlike other civilizations, whose rulers exacted tribute through their control over water, many Mayan farmers were dispersed and somewhat self-sufficient, and Lucero suggests that this led to meaningful differences between Mayan and other civilizations in the way the ruling class exercised power over water. She argues that Tikal prospered not so much because its priestly rulers controlled access to irrigation water for agriculture, but because during the dry months they exchanged drinking and cooking water for labor and fealty (an exchange that was dressed up in the garb of religious obligation). Rather than position themselves as gatekeepers to natural sources of water, the lords of Tikal and other great lowland centers controlled an artificial series of reservoirs.

In the lowland part of Guatemala where Tikal became a center of power, the rains stop in the fall and don't resume until the spring. With no lakes or rivers nearby, people turned to their rulers for water during the dry months. Gill notes that in Tikal the underwater lens of fresh water lies about 425 feet underground—far beyond the 75-foot limit of Mayan well diggers, given the limit of their stone tools. Through a system of reservoirs, however, the nobility provided water to farmers who came in to Tikal during the dry winter season and exchanged labor on palaces and other royal structures for drinking water, which was stored in a variety of natural and artificial sites, ranging from underground chultuns, some of which could supply twenty-five people with a year's worth of household water, to large reservoirs.

The power and the legitimacy of the rulers were based on their pretense that through their rituals and lineage they were conduits to the gods who controlled water. Imagine then the state of mind of these farmers who flocked to Tikal from outlying farmsteads during the winter of A.D. 910 when drought called into question the lords' dicta about their influence with the gods that delivered water. Imagine their mood the next winter when once again, the ruling class's promises to deliver proved as empty as the *aguadas* and reservoirs upon which the city depended. During the next four years, the rains failed as well, by which time, argues Lucero, the lords would have long since lost their hold on the farmers.

One difficulty in isolating the role of climate change in the fall of the Mayans has to do with wide regional variations in climate. If the rains dry up in the lowlands, they might continue in the mountains or in adjacent areas. According to David Hodell of the University of Florida at Gainesville, rainfall varies by a factor of five in the Mayan territory between the southeast lowlands, the mountains, and the arid northwest. Still, in the lowland area around Tikal, climate tended to be relatively stable. The annual south-to-north shift of the Intertropical Convergence Zone (the ITCZ) during the summer months would bring with it humid air and rains that would persist through the late spring to early fall. Indeed, an understanding of when the rains would arrive was crucial to food production. If farmers put maize seed in the ground too early, it would rot, and if they planted too close to the arrival of the dry season, the seeds would fail to germinate.

The Mayans developed a science for predicting the best times to plant during their periods of stability. They also developed an elaborate culture around the collection and distribution of water during the dry months. As Lucero describes it, they learned not only how to impound water, but how to keep it potable. Mayan iconography is filled with the image of water lilies. The nobles were referred to as "water lily people," and an image of one of the Mayan gods consisted of a jaguar's head with a water lily on it. The significance, says Lucero, is that the water lily, particularly the genus that shows up in Mayan iconography—*Nymphaea*—is particularly sensitive to changes in water quality. It requires water that is not too deep, and it can't tolerate a large burden of algae or bacteria (and it provides the service of filtering out pollutants). The water lily thus correlates not just with water, but also with potable clean water. It is little wonder that it rose to iconographic significance in a society that depended utterly on the ability to store and maintain clean water in a hot region that was dry for nearly half the year.

According to Lucero, the system of reservoirs would have enabled the rulers of Tikal to get through one year of drought and perhaps two. But as the drought continued in the years following 910, the theocracy would have been forced to reduce and then renege on promises to deliver their water-for-labor bargain. At this point, those farmers who were relatively mobile and self-sufficient probably began to ask themselves why they should continue to provide labor and part of their agricultural surplus to a theocratic aristocracy that based its legitimacy on its ability to deliver water. "At some point," says Lucero, "the farmers simply said, 'See ya, we're outta here,'" and headed off in search of better conditions. The lords of Tikal fell into a spiral of internal strife and wars with water-blessed neighbors. At the start of the collapse, the Mayan population was probably near its peak, and so the effects of drought would have compounded the loss of crops through environmental degradation and erosion that resulted from deforestation and overplanting during wetter times.

The bare bones of this retelling does not do justice to the horrors that must have accompanied this collapse. While some farmers probably scattered and survived, millions perished, some through conflict, others through disease, and certainly many through thirst or famine. Gill likens drought to HIV in the sense that just as the virus sets up its victims for cancers, TB, and other lethal ailments, drought weakens its victims to the point where, as he puts it, "in most famines it is impossible to tell whether the victim has died of starvation or disease, the two are so intertwined."

The Mayan collapse was one of history's all-too-frequent periods in which the living must have envied the dead. As Gill points out, only war, infection, and famine can account for the millions of people who disappeared during the collapse, and war would have offered the easy way out. During starvation, the path to death can take months in which the body slowly consumes itself. Along the way, the skin changes color and loses its elasticity, while people slowly lose their ability to digest, their morals, and ultimately their minds. During the last stages of starvation, one might gain a fine fur that covers large parts of what remains of one's body. Cannibalism is a recurrent theme of big famines. Depending on their condition at the start of a famine, death might come at different times, but most enter the fatal zone when they have lost between one-third and one-half of their body weight.

Although Mayan civilization had collapsed and reconstituted itself more than once during the previous thousand years, this time it went down for keeps. Six hundred years later, when the Spanish began their conquests in the Americas, all that remained of the Mayans were a few scattered outposts.

Tales of a civilization's collapse cast a retrospective taint of failure on a people's history. Forget about the glories of art or technology a culture might have created, its fall brings into focus moral failings or other flaws that brought about decline. Rome incubated the rise of the West, but most students reviewing the life and times of the empire are apt to look for early signs of the corruption. Similar morality plays of arrogance and decadence punctuate popular appreciation of long-lived societies from Pharaonic Egypt to the Ottomans.

Perhaps this is one reason why historians and archaeologists resist the notion that climate might play a decisive role in the rhythms of history. To the degree that climate contributed to the fall of the Akkadians. Assyrians, or Minoans, the self-appointed gods of these civilizations escape blame. There is a point, however, where the evidence of an external deus ex machina such as climate reaches such a clamor that it's impossible to ignore, no matter how unattractive and unsustainable the practices of the ruling class. That seems to be the case with the Mayans.

7

The Little Ice Age: Five Hundred Years of Climate Chaos

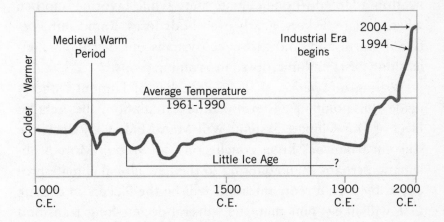

As LONG AS climate remains stable, winners and losers will sort themselves out. The Inuit prospered while the Norse failed during the Little Ice Age in part because the Eskimos could hunt ringed seals from sea ice during the winter, and the expanding sea ice presented them with expanding opportunities (indeed, during the coldest part of the LIA, kayak-paddling Inuit showed up repeatedly in the British Isles). For those Inuit still pursuing traditional ways today, the present warming is a disaster as sea ice retreats (and if the Norse had held out for eight hundred years, they would find a climate in Greenland similar to that when they first arrived). It's not climate, but climate change, that throws civilizations into a tailspin. Critics opposed to taking action to avert climate change have seized on the notion of winners balancing losers in recent years, but a close look at how climate changes is less than reassuring.

Climate does not typically shift from consistently warm to consistently cold. Climate transitions are typified by "flickering," a period when climate rapidly shifts back and forth between warm and cold, wet and dry, before settling into a new state. And, if changing climate imposes hardships on a civilization adapted to one climate state while favoring another, when climate flickers, nearly everybody loses. Rapid shifts between warm and cold throw ecosystems out of balance, unleashing pests and microbes, and ruining crops.

Writing in *Science*, Wallace Broecker of Lamont-Doherty made this point in "Thermohaline Circulation, the Achilles Heel of Our Climate System: Will Man-Made CO_2 Upset the Current Balance?" In his conclusions, he wrote: "More problematic perhaps than adapting to the new global climate produced by such a reorganization will be the flickers in climate that will likely punctuate the several-decade-long transition period." He was writing about potential threats in the future, but there have been many such transitions in the past.

The most detailed recent example of the misery and death that follow what could be described as a flickering climate comes from the years before and during the Little Ice Age. In this case, climate historians have access to rich historical records to supplement broader proxies such as ice cores and tree rings. Tax rolls, church records, and countless other documents enable us to reconstruct the weather in particular districts right down to its effects on individual families. There's still no consensus on what caused the Little Ice Age or even whether it was one continuous event or rather a cluster of climate episodes. Leaving aside the definition of the event, the swings in climate and multiple cold episodes are undeniable. The word "flicker" seems harmless, but the rich historical and climate record makes it possible to show that even a relatively modest episode like the Little Ice Age can produce devastating effects.

Ranked against some of the other climate events of the Holocene—the coolings of 8200 B.P., 5200 B.P., and 4200 B.P., for instance—the overall footprint of the Little Ice Age was slight. Global temperatures dropped between 0.5 and 1 degree centigrade. Even in the North Atlantic region, which bore the brunt of the event, temperatures fell by about 3 degrees centigrade on average according to Peter deMenocal, about the same as during the frigid winter of 1977-78 in the Northeast. That winter, however, was cold enough to choke the Hudson River with ice to the point that icebreakers had to clear a path for ships delivering supplies to Albany, New York. While the cold weather relented in the Northeast after that one winter, during the deepest parts of the LIA, the frigid winters continued year after year, at times for decades, and on and off for more than five hundred years. Moreover, stretches of cold weather were punctuated by occasional warm and wet periods marked by floods and windstorms.

The centuries of wild and harsh weather left their imprint on the peoples of Europe. By the end of the deepest part of the cold—the so-called Maunder Minimum between 1645 and 1715—there were barely as many people in Europe as there had been four hundred years earlier. Moreover, they lived shorter lives, and they were of shorter stature than their medieval ancestors.

On the other hand, it can hardly be said that the LIA set back European civilization, since the Little Ice Age also encompassed the Renaissance and the birth of nation-states. In fact, the Little Ice Age may have directly enabled the genius of some European artists. Lloyd Burckle of Lamont-Doherty and others have argued that the cold temperatures of the Maunder Minimum tightly packed the tree rings of the spruce that Stradivarius used to build the sounding boards of his violins, giving his instruments their uniquely warm sound. Indeed, H. H. Lamb and others have argued that by playing a crucial role in the demise of the medieval era, the Little Ice Age cleared the way for the Enlightenment.

The first frissons of the LIA hit when Europe was booming with unprecedented prosperity. That prosperity was itself related to clement weather, in this case the Medieval Warm Period. Between A.D. 900 and 1300, Europe's population increased by a factor of four, according to the late historian Norman Cantor.* England's population tripled in the thirteenth century. Life expectancy increased as well, according to Lamb, who estimates that by the end of that century the average life span was about forty-eight years. The rich farmlands of central England produced bumper crops; merchants in ordinarily chilly parts of Scotland built castles that rivaled those of the nobility. For most of the warm period, famines and crop failures were distant memories.

In England, that halcyon period came to a sudden end in the early part of the fourteenth century (the weather had already begun turning squirrelly farther north a century earlier). The Thames froze in the winter of 1309, and then in 1315 the British Isles warmed up again, but the mild weather was accompanied by a succession of epic storms. More civilizations have been done in by drought than by deluge, but too much rain can be just as ruinous. With the British rains, trees had a glorious time, but waterlogged fields, premature germination,

*Norman Cantor. *In the Wake of the Plague: The Black Death and the World It Made* (Free Press, 2001).

failure to ripen, and rot both above and below the ground, ruined and reduced crops. As the rains extended into the summer months, farmers lacked the sunshine necessary to produce the salt they needed to preserve meat for the winter. The dank conditions turned fields and houses into incubators for molds, pests, and disease.

Livestock as well as crops suffered from the wet and disease. Many cattle and sheep that had survived various plagues froze to death when cold weather returned with a vengeance in 1317. After eating their livestock, including animals felled by disease, villagers reportedly began eating cats, dogs, and in some cases, each other.

Ordinarily, the peasant of the Middle Ages could call on friends and family in hard times, but throughout the British Isles everybody was in the same boat, as were the farmers across the Channel in Europe, not that cross-border trade in crops was sufficiently developed to meet the needs of the starving. As the rains continued through the next two growing seasons, famine and disease began to take their toll not only on the economy but also on the fabric of society. Robbery increased, farmers lost their property, and bands of starving homeless roamed the countryside besieging priories and other storehouses rumored to maintain grain.

In *Climate, History and the Modern World*, H. H. Lamb describes the effects of St. Anthony's fire, one of the diseases associated with the dank conditions. The disease, he writes, was "produced by the ergot blight (*Claviceps purpurea*) which blackened the kernels of the rye in damp harvests. Even a minute proportion of the poisoned grains, baked in bread, would cause the disease. The course of the epidemics was such that the whole population of a village would suffer convulsions, hallucinations, gangrene rotting the extremities of the body, and death. In the chronic stage of the disease, the extremities developed first an icy feeling, then a burning sensation; the limbs went dark as if burnt, shriveled, and finally dropped off."*

Terrible as it was, St. Anthony's fire was mild compared to the pestilence that was brewing in central Asia. The world had already suffered through the epidemiological effects of a swift succession of floods and droughts when the Justinian Plague broke out of its reservoir in sub-Saharan Africa in the sixth century. In 1332, a similar combination of factors made Asia a petri dish for the next iteration of the plague, the Black Death. H. H. Lamb calls the flooding in China's river valleys "one of the greatest weather-related disasters ever known," since the floods led to the deaths of roughly 7 million people. The years following saw severe drought, setting up the climate seesaw that would cause the rapid increase and collapse of various rodent populations, both of which could have brought plague into contact with humans (microbes often jump to new animals when their host's numbers collapse).

At the time of the plague, no one knew what agent carried the disease (or even that agents carried diseases), and so it is very difficult to reconstruct the exact role the weather played in releasing the Black Death from China and Mongolia, where it had been bottled up in rodent populations. Before it made its way down the Silk Road to Crimea, the plague killed an estimated 35 million people in China. Then, in about 1346, it began to move west.

One narrative of the plague's westward expansion has the fleas traveling in the pelts of marmots (woodchucklike creatures) that had been killed by the disease. Trappers, thinking themselves lucky, picked up the dead marmots, skinned them, and sold the pelts to dealers who then carried them and their ravenous fleas along trade routes. Supporting this theory are spotty contemporary accounts of a horrible disease decimat-

*H. H. Lamb. *Climate, History and the Modern World*, 2d Ed. (Routledge, 2002).

ing trappers in the Asian steppes. The fact that the marmots died of the plague indicates that they were not a reservoir for the disease, and this in turn suggests that they became infected either as the natural hosts died off (because of drought) or through contact with the host rodents in the aftermath of the floods. In any event, according to this theory, once the plague reached the ancient port of Kaffa (now Feodosiya) in Crimea, the fleas jumped from dead marmot skins to black rats, which put them on the fast track to spread to humans. After initial reports of deaths in Kaffa (the disease killed in just a few days), the Genoese tried to quarantine arriving boats, but the rats simply swam ashore, and the great pandemic began.

No one really knows how many people the plague killed in Europe. Estimates vary from between a quarter and one half of the population, with absolute numbers ranging from about 20-to-50 million people. Already weakened by famines and other diseases, the population had no resistance to the plague, and it simply tore European society asunder. Cities became de-populated as people died or fled to the countryside (Paris was cut to half its preplague size); landowners found themselves without manpower to work the fields; Jews were blamed and slaughtered from Russia to the British Isles. Societal norms broke down: troupes of disoriented penitents took to display-ing perverted sexual acts in public, a symptom of the death throes of a social order. Science could not halt the disease nor could clerics offer solace.

The plague and other epidemics made several return visits over the next few centuries. All these traumas, direct and indi-rect effects of the LIA, left their imprint on the demography of Europe, and some of the reverberations continue to this day. The last decade of the sixteenth century, for instance, wit-nessed an intense cold spell. Drawing on records and chroni-cles compiled by others, Lamb shows that in Scotland four of the ten years of the decade witnessed local or general dearths, and famine struck in two other years. When crops failed again in 1612 because of the weather, King James VI kicked many Irish out of Ulster, which was slightly less affected by the savage weather, and allowed Scottish farmers to move in. By the end of the century, 100,000 Scots had established themselves, setting the stage for the religious conflict between Protestant and Catholic that has dogged the region for the past three hundred years.

The weather took its toll elsewhere in the world as well. Long after the LIA ended the Norse experiment in the New World, the intermittent cold and drought of the period continued to wreak havoc. In North America, where the British were establishing their first colonies in the late sixteenth century, one of the early failed attempts involved the so-called Lost Colony in Roanoke on what are now North Carolina's Outer Banks. Based on scouting reports sent back by a 1584 expedition, Sir Walter Raleigh financed an initial colony in 1585. Raleigh's cousin, Sir Richard Grenville, delivered the colonists, and Ralph Lane was left in charge as governor of the newly claimed territory. The colony never quite became self-sufficient, and when Sir Francis Drake's efforts to resupply the colony were thwarted by storms, he decided to evacuate the colonists back to England in 1586. Unmindful of Drake's rescue, Grenville sent a supply ship shortly thereafter. Finding no one, the ship left fifteen men to hold the fort and returned to England.

Undaunted, Raleigh organized a larger colonial group the next year, sending 150 men, women, and children to Roanoke. The terms would probably attract colonists even today since Raleigh offered a minimum of 500 acres to each family, and more if they could contribute materially to the expedition. Governor John White established the colony on July 22, 1587, and then in late August returned to England to secure more supplies. This was the last contact that any known European had with the colonists. For a variety of reasons, White's efforts at resupply did not come to fruition until 1591. When he did return, he found an abandoned settlement. He came upon the word "Croatoan" carved into a palisade, but none of the other symbols of a pre-arranged code that would have signaled an emergency evacuation. Thinking that the group had moved to another island under the control of the friendly Croatoan Indians, he started off in that direction in the hopes of finding his daughter and Virginia Dare, his granddaughter, the first English baby born in North America. He was running out of water, however, and as he headed south to the West Indies to replenish his supplies, a storm blew his ship all the way to the Azores. The fate of his family notwithstanding, White sailed north to England.

Raleigh did not give up so easily, although this was in part for financial reasons. A brief imprisonment (the queen felt he exhibited bad manners by marrying without her approval) had voided his rights to Virginia, but he felt that as long as his colonists were still living there he could assert his prior claim. After he got out of prison, he sent off an expedition to find his colonists. They too came back empty-handed (and some historians question how hard they tried), and the enduring mystery of the "Lost Colony" was born.

For many years, historians focused on hostile natives as the likely culprits, and indeed, periodic rampages and slaughters by the British certainly gave the local tribes sufficient motive for revenge. In 2003, a group of paleoclimatologists led by D. W. Stahle of the University of Arkansas's Tree-Ring Laboratory reconstructed a climate history for the region going back to A.D. 1185. Looking at a continuous series of rings taken from bald cypress trees, they discovered that the worst drought in an eight-hundred-year span started at virtually the same time that the colonists landed. With no water and no way to get it, the colonists faced long odds on the Outer Banks.

Then there is Jamestown, whose fate two decades later somewhat bolsters the hypothesis that water played a roll in

Roanoke's downfall. Situated on the mainland, the Jamestown colonists had more options in securing water supplies, yet nearly half its residents died, many as a result of malnutrition, during an intense six-year drought that lasted from 1606 to 1612.

In the later decades of the eighteenth century, the LIA was still causing trouble in the colonies. In 1765, the British garrisoned 6,000 troops in the Northeast and required that settlers provide them with food. This came precisely at the time when colonists were having trouble meeting their own needs following a series of killing frosts, according to the historians David Smith and William Baron. Smith and Baron write that the burden added to the incendiary mood in the colonies, which erupted into open revolt eleven years later.

The weather did not stop the European settlement of America. It did set back the timetable for permanent settlements, possibly by more than six hundred years if we start the clock with the first Norse longhouses. The LIA also tilted the wheel in favor of the more southerly countries. While sea ice virtually put the Norse out of business, the frequent violent storms of the era merely slowed the pace of exploration by more southerly Europeans such as Venetians, Portuguese, Spanish, and English adventurers who took to the seas below the ice limit. In fact, Lamb argues that in the fifteenth century some portion of English, French, and Portuguese maritime exploration was motivated by the need to follow the cod stocks, which had been forced south by the increasingly frigid waters in the northern ocean.

So what lesson might be taken from the Little Ice Age? Had the benign weather of the Medieval Warm Period continued for another five hundred years, would history have been different? Most certainly: The Norse might have taken greater advantage of their head start in the New World. The population explosion of the eleventh and twelfth centuries would have likely continued for some time in mainland Europe, prolonging the manorial society and depressing the bargaining power of peasants. The history of humanity, however, is a succession of societies growing to fill every available space and then some, and it is unlikely that Europe would have evolved into a utopia of peace and love. The exploding numbers and wealth gap of feudal society would eventually have brought about a collapse of the old order, either through pestilence (crowding can destabilize ecosystems as much as the weather), ecological collapse, corruption, conflict, or, most likely, some combination of all of these factors.

The Little Ice Age metered out its challenges and provided an example of the motto of athletes and soldiers: "That which doesn't kill you makes you stronger" (a truism exemplified at the evolutionary level during the glacial age climate challenges that crafted modern humans). The privations of the period provided an impetus for European royalty to seek lands and wealth beyond the borders of the continent, and, as Alfred Crosby has argued in *Ecological Imperialism,* the survivors of its various pandemics went forth into the world with biological weapons and an immune system that together comprised a weapon of conquest (and was consciously used by both the Spanish and the English).* It's true that Europe ultimately met the challenges of the Little Ice Age, but the triumph was over a relatively minor event in the great scheme of climate change.

The preceding chapters present a case for including climate change as a factor in human affairs. As with any case, the jury wants to know how reliable is the forensics that produced the evidence, how reliable is the science behind the forensics. It is to these questions that we now turn.

*Alfred W. Crosby. *Ecological Imperialism: The Biological Expansion of Europe 900–1900* (Cambridge University Press, 1993).

PART TWO

Evidence

8

Climate Comes into Focus

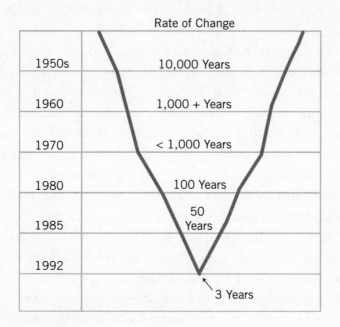

Rate of Change

HISTORIANS ARE ALWAYS on the lookout for some new idea with the power to explain. Over the years, diet, desertification, and deforestation have appeared as unappreciated forces in history, not to mention the factors of *Guns, Germs, and Steel,* offered by Jared Diamond.* There is a typical pattern to these vogues. In the first phase, a historian or scientist argues

*Jared Diamond. *Guns, Germs, and Steel: The Fates of Human Societies* (Norton, 1997).

[8 9]

that some factor has not received proper recognition for its role in human affairs. The academic-scientific immune system at first rejects the new idea, but eventually resistance gives way and this new idea spreads as legions of grad students reinterpret the past in terms of this new player. Inevitably, in a corollary of the Peter Principle, the idea is pushed beyond its proper realm. As exceptions and anomalies mount, enthusiasm wanes, and academia moves on to embrace some new explanation. Over time, though, if the factor has merit, it finds its place.

So it was when the University of Chicago historian William H. McNeill elevated disease as a force in history in his book *Plagues and People* in 1976.* Long before McNeill, scholars had documented the ways in which smallpox had weakened the Aztec Empire, but McNeill argued that the disease was the Spaniard's decisive weapon, and that at many points in history plagues have had more impact on the fortunes of societies than conventional factors such as innovation and resources. His book prompted a flood of revisionist retrospectives. Ten years later, the distinguished historian Alfred Crosby published his *Ecological Imperialism,* in which he explored the ontogeny of the West European immune system as an instrument of conquest that permitted imperialists to prevail over a panoply of cultures in far corners of the globe. More recently, Jared Diamond has taken disease as a force in history and placed it in the context of other factors, such as environmental degradation, technology, and weaponry.

The argument that geography and climate determine economic development dates to the eighteenth century. In 1776, Adam Smith argued in *The Wealth of Nations* that the economic potential of nations could be measured by assessing access to navigable rivers (which provided access to resources in

*William H. McNeill. *Plagues and People,* updated ed. (Anchor, 1998).

a continent's interior) and to protected ocean ports.* In the early twentieth century, Ellsworth Huntington of Yale promoted a corollary of this idea, called economic geography, and took it further to explore climatic determinism as well.

One postulate of economic geography back then held that more than two-thirds of differences in income among nations could be explained by distance from the equator: the farther away, the richer and more developed the nation. Called "the equatorial paradox," this rule of thumb was widely derided as racist since it implied that contending with winter required people to be more industrious (as embodied in the old banker's dictum: "Never lend where gentlemen don't wear overcoats in the winter"). Unfortunately, Huntington fueled this interpretation of climatic determinism by freely speculating that the crisp air of the northern climes created the conditions for clearer thought and creativity.

Lost in the backlash to such claims were other findings linking climate and human evolution. The renowned paleontologist Raymond Dart published a paper in *Nature* in 1925 describing a homonid skull and concluding with the argument that the savannah climate of southern Africa with its "relative scarcity of water, together with a fierce and bitter mammalian competition, furnished a laboratory such as was essential to this penultimate phase of human evolution." Instead of following up on Dart's speculations, the archaeological community turned away from linkages between environment and human evolution to the degree that for much of the twentieth century published papers almost never mentioned paleoclimate when discussing human evolution.

For decades, the field of environmental determinism languished in the purgatory of the politically incorrect. While it was impolitic to talk about the environment as a shaper of

*Adam Smith. *The Wealth of Nations* (Bantam Classics, 2003).

brains and civilization, however, it has gradually become acceptable to write about environmental factors as destroyers. Take, for instance, deforestation as a force in history. John Perlin's marvelous book *A Forest Journey* explores the correlations between the destruction of forests and the decline of civilizations down through the ages.* In Perlin's view, unchecked human appetites play a major role in the episodes of deforestation that contributed to societal collapses from the Hittites and Babylonians and on down through history. In several of these cases, however, human folly may have been amplified by another factor: climate.

The revival of interest in climate as a shaper of humans and history came not from the archaeological community, but from chemistry and genetics. In the 1980s, Elisabeth Vrba, a Yale paleontologist looked directly at the role of climate in human evolution. She had the common sense to wonder what was happening to other animals when human ancestors underwent rapid change. Studying fossil pollens, she found that at selected points when human ancestors were undergoing rapid evolutionary change, many other mammals were changing as well. Her initial papers published in the 1980s set the stage for Potts's work on climate and human evolution discussed in Chapter 2.

The idea of linking human evolution and climate change is just another episode in an ongoing reevaluation of the role of climate in human affairs. The idea's recent provenance (twenty-five years or so) is not terribly surprising, since the two theories that gave rise to this linkage are not very old themselves. Evolution dates back to the mid nineteenth century, of course, and it was just a little earlier, in 1837, that Louis Agassiz put forth the notion that ice sheets once covered much of Europe. The reaction to this earlier suggestion was

*John Perlin. *A Forest Journey: The Role of Wood in the Development of Civilization* (Norton, 1989).

only slightly more polite than the jeers and mockery that greeted Darwin (perhaps most eloquently stated by the British statesman Benjamin Disraeli, who testified that if there was a question of whether we were descended from apes or angels, he was on the side of the angels).

Further evidence that climate has shed its earlier taint came from a study inaugurated by economist Jeffrey Sachs when he ran the Center for International Development at Harvard (Sachs has since moved to Columbia University), which put to the test Adam Smith's original idea. Despite the massive changes in transportation and technology since 1776, he discovered that using Smith's criteria, you could have predicted that countries like the United States and most of Western Europe would do well in the years since Adam Smith offered his thoughts. On the other hand, sub-Saharan Africa's economic prospects are limited according to Smith's criteria. Much of the continent is perched up on a shelf, limiting river access to the interior from the sea. Confirming another aspect of economic geography, Sachs's data documented that workers are demonstrably more productive in northern countries where freezing winters kill off mosquitoes and other disease-bearing insects, while African economies pay a huge penalty as infectious diseases sap energy and reduce work time.

The real momentum behind climate's rehabilitation has been the extraordinary advance in the precise dating of past climate events. Fueled by concerns about global warming, geochemists, physical oceanographers, and geologists have made breathtaking advances in just the past decade. As an example, consider the ever more precise picture of past Mesoamerican climates that has emerged in the past decade.

In 1997, David Hodell of the University of Florida published a study of lake-bed sediments from the Yucatán Peninsula that showed a broad band of aridity between A.D. 800 and 900. This suggested that drought played a role in the

Mayan decline but was not specific enough to link the drying to particular historical events. Then, in March 2003, *Science* published the results of an investigation by Gerald Haug, the head of the Department of Climate Dynamics and Sediments group at the GFZ Institute in Potsdam, Germany, that examined the more undisturbed seabed sediments from the Cariaco Basin off Venezuela.

Extracting a slab of mud, Haug's team ran the sample under an X-ray fluorescence unit that had been tuned to read the exceedingly thin laminations in the sediments, and they were able to see the chemistry so precisely that they resolved the picture of climate to the level of seasonal change. The varves (annually deposited layers of sediment) revealed that regional climate entered a 150-year period of recurrent droughts during the period of decline. The study showed that between A.D. 800 and 900, the Mayans suffered through three different droughts of three, six, and nine years. The dates, A.D. 810, 860, and 910 correlated with specific episodes in the collapse of the Mayan civilization.

With Haug's discovery, it's now possible to show how drought played a role in other factors that historians had previously pointed to when analyzing the end of Mayan power, particularly since the Mayans entered the period of drought with a bulging population, setting the stage for famine as forests burned and water evaporated under the incessant sun.

There are other reasons that help explain why archaeologists have tended to downplay climate as a historical force. We, as a species, have tended to look at climate as immutable. As Yale archaeologist Harvey Weiss puts it, "Most historians had the notion that climate was static, and that populations would adapt to static conditions." Raymond Dart made this very statement in his 1925 paper, noting that in Southern Africa "climatic conditions have fluctuated very little since Cretaceous times."

It can be argued that if climate were immutable we humans would not be here to have any opinion on it. We may owe our existence to climate at its most mutable—the most violent climate shifts since the cataclysm that killed off the dinosaurs. Climate may be a serial killer, but it also forces open the door of ecological opportunity. Evolution would produce a very different cast of characters in a world where the climate never varied.

Given the degree that climate has dominated if not determined the fortunes of various civilizations, it is remarkable how little western science knew about how climate changes before the recent past. Perhaps a partial explanation might be found in the difference between climate and weather. Weather is what we experience on a day-to-day basis, while climate describes the longer-term patterns that govern the seasons of a particular region. From the dawn of organized societies, humans have developed ways of anticipating the changes of seasons as well as regularly recurring weather events. Nothing surprising here, since survival, not to mention power and wealth, usually derived from agricultural production. Few early civilizations, however, were around long enough to develop tools to probe past weather, much less delve into the many cycles that change climate. The scientific understanding of larger climate cycles and how climate changes is a relatively recent phenomenon. It was only in the early 1990s with the core samples taken from Greenland that scientists began to realize the degree to which abrupt and wild swings characterize past climates. Abrupt and wild swings buffet humans, to be sure, but also the plants and animals upon which we depend.

This points the way to understanding how climate leaves its imprint: in a number of cases where disease or deforestation has played a role in the fate of civilizations, changes in climate have set the stage for disease or deforestation. Climate plays a crucial role in access to ports and navigable rivers, two of Adam Smith's enabling factors in economic development. Sea ice severely reduced European access to Iceland for over four hundred years during the Little Ice Age.

If disease, deforestation, and geography have been factors in historical events, then climate has been a metafactor, influencing other factors that play a role in favoring or penalizing nations and peoples. Climate does not control geography, of course, but climate can overrule the advantages that geography might otherwise confer. Climate is not the only factor in the spread of disease, but it can create a context favorable to infectious agents, particularly when it changes.

The role of the Federal Reserve chairman has been likened to that of a host who takes away the punchbowl just as the party is getting to be fun. The same could be said for climate change. A tribe or clan moves into a new area, cultivates the land to the limit of its technology, and the population expands. Times are good and the tribe multiplies, eventually occupying every available space. Then climate throws one of its periodic curveballs and growing seasons shorten or rivers dry up or pests are unleashed or all of the above.

This simple pattern exposes the fallacy of the idea that our forebears had the luxury of moving to vast open spaces when things turned bad in one area (although it's true that colonial powers have been able to treat conquered territories as a vent for surplus by simply shoving aside those already there). Even though human numbers have exploded since prehistory, at almost every point along the way, when climate was stable, human population quickly rose to push the limits of the carrying capacity of the land, given the technologies and mobility of the times. Whenever climate conditions have permitted human numbers to expand, our species has expanded to and beyond the limits imposed by a hunting-and-gathering way of life, or early agriculture, or irrigation, right up to the present.

This proclivity to procreate has periodically left cultures prone to crashes when climate has changed. As Peter deMenocal puts it: "If you look at these ancient cultures, you see sophisticated adaptations to marginal environments. They developed mechanisms—grain storage, water storage—to buffer themselves against changes, and they did so in ways comparable to modern cultures. But the one thing they can't prepare for is the thing they don't know."

This pattern has also created one of the paradoxes of humans and climate: even though our species survived the cataclysmic events of the ice age, and even though our technologies and social organization have become ever more sophisticated, we have become more vulnerable to ever smaller disruptions of climate (it's not that earlier people were better adapted, but rather that there were fewer of them, they had less to lose, and, perhaps most pertinently, didn't have the tools to leave a record of their suffering). Which is why even the little events of the otherwise benign Holocene have had outsized consequences for those in the wrong time and place.

Because the understanding of climate gyrations is a very recent development, it is understandable that the idea that climate is a factor in history has met resistance. It doesn't help matters that historians are being asked to accept the seemingly contradictory notions that climate instability was good for humans as a species, but bad for civilizations; and that climate stability set the stage for the emergence of civilization, but it would not have led to the evolution of humans in the first place.

And now, of course, we are also being asked to accept that humans can change climate, that climate is now changing, and that we are causing the changes. Just as people get used to the idea that humans might be causing global warming, the idea arises that global warming might cause a sudden freeze. To cap the whole thing off, despite all the assertions about what climate has done in the past, what we are doing to climate, and what it might do to us in the future, it's still difficult to predict the weather more than a few days out. Today, thanks to discoveries about past climate, and how climate changes, we are being asked to accept a lot.

No wonder we are confused. Still, we have a stake in the question of whether climate will reshuffle the deck to the benefit of some creature other than *Homo sapiens*—say, for instance, heat-tolerant microbes or cold-tolerant fish. The climate gods have no more stake in the possibility that fish might take over the world than they did in the ascent of man. And so, continuing the profile of this god/serial killer/giver of life, here follows a brief tour of the climate system, how it works, how climate has changed in the past, what those changes meant for the human family tree, and where we are now.

9

The Gears of Global Climate

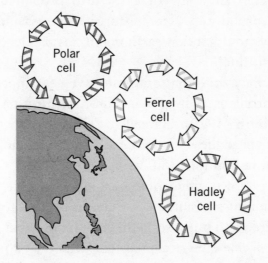

ONE SET OF ATMOSPHERIC CLIMATE GEARS

AT A Rockefeller Brothers Foundation conference in Pocantico Hills, New York, in 2004, scientists, activists, and media and advertising people concerned with environmental issues were each asked to name the most effective piece of environmental advocacy of the modern environmental movement. Without hesitation, I cited the images of the earth that were beamed back in the early days of space exploration. Though the impetus of the space race had nothing to do with environment, for the first time, humanity had an opportunity to view the planet whole, and see the vast oceans, air currents, and weather systems that tie the inhabitants of the planet into one system. Glimpsing the vulnerability of life on earth prompted a powerful protective response from millions of people.

Decades after these first images, I think it will still be years before we realize the true impact of seeing the home planet from space. For the first time, answers lay before scientist's eyes about some of the linkages that connect the land, seas, and sky. The new view of earth also prompted questions. Given the brutal and extreme forces that assault the planet continually, how is it that earth maintains the delicate balance necessary for life?

Those images and that question gave new impetus to a hypothesis first posed by an inventor-scientist named James Lovelock in the 1970s. Named the Gaia hypothesis after the ancient earth goddess of the Greeks, the idea holds that the planet is alive and functions as a superorganism in which living things interact with geophysical and chemical processes to maintain conditions suitable for life. In his first formulations, Lovelock imputed purpose to the interactions that maintained life on earth. He later accepted that even the exquisitely precise balances of the atmosphere (oxygen levels have remained at roughly 21 percent for 200 million years—if that percentage rose to 25 percent, fires would spontaneously break out; if it dropped below 15 percent most mammals would suffocate) might be perpetuated and stabilized by simple feedback. In a hypothetical scenario, Lovelock showed that a planet covered by light- and dark-colored daisies could control the heat received from the sun. In this self-regulating model, dark daisies would absorb sunlight and warm the planet, until it became too hot for the dark daisies and instead favored the proliferation of light-reflecting daisies. That would have the effect of cooling the planet until the cycle reversed itself again.

Regardless of whether earth amounts to a superorganism,

the Gaia hypothesis offers a helpful model for envisioning the intricate interconnections that maintain and balance the climate system. Indeed, perhaps the biggest beneficiaries of space imagery were geophysicists and other climate specialists who could now see whole climate systems that previously had to be pieced together from data collected around the world. The tremendous bounty of a new perspective on global climate was as humbling as it was invigorating. Here before scientists was a system in which everything, from earth's position in its orbit around the sun to what's growing on the ground, influences climate. How the climate system balances these various inputs and feedbacks is a problem as complex as life itself.

Let us take a brief tour through the basic climate system. The tour starts with a description of the way the spinning globe distributes the sun's heat. Then it moves into the cosmic and planetary events that change climate over long timescales, and subsequently describes some of the shorter cycles that ripple through climate in response to these big events. The tour concludes with a description of where we are now in the interplay of climate cycles, big and small.

The basic engine of climate is the sun, which delivers a reasonably steady amount of energy to the earth each day. (That energy changes over time though on very long timescales.) Think of it as an allowance given daily to the planet, mostly at the equator. The amount of energy and its distribution vary according to factors that affect temperature, precipitation, and wind on timescales ranging from hundreds of millions of years to the change of seasons. Through blind physics and biophysical and geochemical feedback, the system maintains relative stability even as one or another variable changes. Some of these key factors include variations in earth's orbit, the spinning of the globe, movement of the continents and the resulting array of oceans and landmasses, the shape of the ocean floor, mountain ranges on land, sunspot activity, the composition of the atmosphere, the waxing and waning of ice sheets and glaciers, the color of land and water, as well as extraordinary events such as direct hits from giant meteors or episodes of volcanism. What is living on the planet can have big effects as well. Deforestation, for instance, can have large effects on regional climate and can impact rainfall patterns far away from the affected forests. Picking one's way through the interplay of climate cycles and the biosphere is like entering a wilderness of mirrors. It's hard enough to identify the various forces and conditions that impinge on climate; sorting out the myriad interconnections of these factors introduces a new level of complexity.

Today about 30 percent of the sun's incoming radiation bounces off light-colored surfaces such as ice, snow, clouds, and sand, and is reflected back to space without heating the earth (obviously, the amount reflected can vary with changes on the planet that affect the reflective surfaces of the planet). Overall, the energy budget of the planet finds a balance. If more energy came in than was used on earth or radiated back to space, the planet would steadily warm until that budget came into balance. Conversely, if the energy used on earth or radiated back to space exceeds the amount coming in, the planet would steadily cool.

Of the energy that gets through, a good portion of the daily allowance of the sun's energy enters near the equator, where light rays hit the earth most directly. (The tropics can't get too hot because if sunlight becomes stronger, evaporation simply makes more clouds, which reflect heat before it hits the surface.) Were the earth not spinning, this energy would pile up, and, following the laws of physics, gradually spread north and south from the equator, rising in the atmosphere and spreading through water vapor in air currents, while fingers of warm water snaked toward the poles in the oceans. The world is spinning of course, which complicates this dry exercise in energy flows.

As the globe spins on its daily rotation, the atmosphere that cloaks the globe spins with it. If it didn't, winds at the equator would be 1,000 mph as the earth hurtled through the still air. In Stockholm and Anchorage, however, these winds would be about half as fast, and at the poles, there would be no wind at all. The spinning, and its different speeds at different latitudes, have large effects on the capture and distribution of the sun's energy as it warms land and water.

Imagine a child standing on the equator with a very large squirt gun pointed due north. When he squeezes the trigger, he launches a jet of water northward and, at first, as the stream of water leaves the gun, it heads due north. But as this jet is moving north, the earth is rotating toward the west underneath the water so that the path the water takes over the land seems to curve toward the right. That's because the only force impelling the water was the initial squeeze of the trigger. The only way the child could make the squirt move straight north would be to find a way to exert constant westward pressure on the jet to compensate for the earth's rotation. Otherwise, as the water moves north, its path trails off to the east as the initial force dissipates in response to friction and gravity. If he turned around and squirted his gun due south, the same thing would happen; only the stream would trail off to the left.

This apparent force is called the Coriolis effect or force, named for Gustave-Gaspard Coriolis, a French mathematician who first described it in 1835. It has major effects on climate. As heated air rises over the equator and begins spilling north, the Coriolis effect diverts its movement toward the east. Along its route, the air cools and begins sinking in the horse latitudes between 25 and 30 degrees north and south (so named because sailing ships would often become becalmed in these areas of light winds and the crew would sometimes throw horses overboard to lighten the load). The sinking air is drawn back toward the equator by the low pressure created there by rising heated air. As this air flows south and west, drawn back in as the rising air creates a partial vacuum, it forms the trade winds, so named because trading captains relied on these constant winds.

The whole loop, consisting of air rising in the tropics, sinking in the doldrums, and then flowing back toward the tropics, is called a Hadley cell. An idealized diagram of this cell as a vertical slice of the atmosphere would look like one big gear rotating counterclockwise. Horizontally, it would extend between the equator and 30 degrees north and vertically between sea level and 50,000 feet. North of the Hadley cell, neatly occupying the latitudes between 30 and 60 degrees north, would be another giant gear called a Ferrel cell. The Ferrel cell lies between the Hadley cell and a third cell that lies from 60 to 90 degrees north. This Polar cell circulates warm air northward that then cools and sinks at the poles to return southward at the surface. Caught between these two gears, the Ferrel cell circulates air toward the north at the surface, rises where it meets the Polar cell, and then returns air southward at high altitudes. The Coriolis effect bends this surface northward flow to the east, giving the northern hemisphere its characteristic westerlies. The southern hemisphere mirrors this three-cell system.

Driven by heat collecting at the equator and the spinning of the earth, this collection of atmospheric gears moves about half the excess heat collected at the poles toward the northern latitudes. Were it not for the Coriolis effect, which diverts this flow eastward, the poles would be much warmer (this is the situation on Venus, where the Coriolis effect is much weaker because that planet rotates very slowly). About 50 percent of this atmospheric transport of heat comes in the form of storms, which draw in warm, moist southern air and then move north. Storms carry much of that heat in water vapor and then release the energy as the water vapor condenses in the colder air. It takes 540 calories to evaporate a gram of water, for instance. That evaporated water travels until it hits cooler air, which has less capacity to hold water vapor. As the moisture condenses into droplets, a good portion of those 540 calories returns to the atmosphere, warming the air. Hurricanes can move massive amounts of heat, cooling the tropical waters and warming the air by releasing heat as they weaken while moving north.

The other big heat distributors on the planet are ocean currents—such as the Gulf Stream—that move the other half of the equator's excess heat toward the poles. The atmosphere and ocean are entwined, each constantly affecting the other. Oceanographers and atmospheric scientists use the phrase "coupled system" to describe the interplay of two of the dominant factors in the earth's climate. It's a system that distributes heat in three dimensions. Everywhere, when warm air is rising, cooler air is being pulled in to replace the rising air. Similarly, in the oceans, cool water surfaces to replace warmer water driven by winds along the surface.

The trade winds drive warm Atlantic Ocean water westward along the equator, where it eventually piles up when it encounters the geographical barrier of Panama (more on this later). This movement of surface water has two major effects on climate. On the one hand, as the surface water is blown westward, an upwelling of cooler, nutrient-rich deep water rises to replace it. Thus, just to the north and south of the equator in the Atlantic and Pacific are two zones of cool waters in spite of the fact that they receive the sun's most intense energy. The wind-driven warm water on either side of the equator eventually spills north or south, guided by the winds and the contours of the ocean basin.

The warm water spills north and south through a series of ocean currents that mirror to some degree the cells in the atmosphere above the water. Like the atmospheric cells, for instance, these giant oceanic gears—called gyres—fall under the influence of the winds and the Coriolis effect and form loops that move water first away from the equator, then parallel, and then back to rejoin the equatorial currents. The familiar Gulf Stream is one of these gyres, and because of its unique characteristics it has particular salience to the fortunes of many civilizations.

Alone among the equatorial gyres north of the equator, the Gulf Stream has extensions that deliver heat very far from the equator. The North Atlantic Current and North Atlantic Drift drag heat much farther north than any Pacific Ocean current. Thanks to this peculiarity of the Atlantic currents, temperatures along the Norwegian coast are about 25 degrees Fahrenheit warmer than places such as Nome, Alaska, that lie at equivalent latitudes. This relative warmth helps explain why European peoples developed material cultures far more elaborate than tribes inhabiting equivalent latitudes elsewhere: simply put, it's hard to build a civilization on lichens.

The Gulf Stream collects its warm water from equatorial currents that flow up from the Yucatán Channel through the Florida Straits and in from the north of the Windward Islands. Then, driven by southwesterlies and corralled by the continental shelf, the warm water heads northeast across the Atlantic. Most of these waters turn eastward and then head south to rejoin the equatorial flow, but a portion splits off and wanders up past the British Isles, Iceland, and Scandinavia.

The Gulf Stream is, in essence, a gigantic hot river moving through the Atlantic. It has been pumping a portion of its warm water to the far north for at least 2.5 million years. More than 60 miles wide in many places, and with an average depth of 3,300 feet, the Gulf Stream books along, moving as fast as five knots. It moves staggering amounts of energy. Water weighs 773 times as much as air, which means that a relatively small amount of water can store a huge amount of heat. There's as much heat in a cubic meter of water at the surface of the ocean as there is in the entire seven-mile-high column of air above it.

In an idealized climate system on a static and uniform globe, the Gulf Stream would be just another giant ocean gyre, and climate would be pretty predictable, and vary gradually and predictably as earth wobbled on its axis and moved through its orbit around the sun. But the globe is not static and its surface is not uniform. In fact, the peculiarity of the Gulf Stream that generously delivers heat to the far north is partly the result of one of the biggest and longest influences on climate.

If a habitable northern Europe is in part a gift of the Gulf Stream, the current's northward wanderings represent, in turn, a gift of continental drift. Of all the variables that can change climate, the position of the major landmasses is probably the longest cycle of all—on the order of 100 million years— and the shifting of landmasses over time has played an enormous role in creating the perfect climate for humanity. On these timescales the present path of the Gulf Stream is one of the more recent developments.

As noted in the first chapter, the Gulf Stream and its northern limb comprise crucial portions of what Wallace Broecker described as the Global Ocean Conveyor Belt (sometimes called the Great Ocean Conveyor), a continuous loop of deep ocean and surface currents that distribute enormous amounts of heat during its 1,000-year cycle. Before there could be a conveyor (at least in its present form), a number of pieces had to fall into place. One of them was the formation of ice sheets in Antarctica, and here too plate tectonics played a role.

The growth of the first ice sheets in Antarctica may have resulted in part from tectonic shifts that isolated the continent. A belt of globe-circling winds formed a polar vortex, which acted as a barrier to storms in the midlatitudes that might deliver heat to the continent. Nurtured by this thermal isolation (there are various theories for the specific mechanism that built the ice sheets, ranging from a shift in ocean currents from warm to cold to a precipitous 80 percent drop in CO_2 in the atmosphere), ice sheets grew up to 3 miles thick (15,600 feet to be exact), entombing 70 percent of the world's fresh water and lowering sea level several hundred feet in the process.

It's hard to imagine the scale of the Antarctic Ice Sheets. When I flew from McMurdo Station to the South Pole in 1997, our route took us along the Transantarctic Mountains. Many of these peaks rise over 14,000 feet—as tall as the highest peaks in the lower forty-eight states. I could see the scale of these mountains as they rose above the Ross Ice Shelf, but on the other side of the mountains an endless sea of ice, starting just below the peaks as though the mountain range was the rim of a bathtub, extended for over 2,000 miles.

The changes that brought about the growth of ice sheets had other effects as well. In Antarctica, the seasonal appearance of sea ice every winter more than doubles the size of the continent. This has its own profound effects on climate. Instead of sunlight hitting dark, heat-absorbing seawater and land vegetation, it now bounces off the brilliant white of ice and snow. Apart from reducing the amount of energy captured by the oceans and atmosphere, the ice regime of the southern ocean also provides one crucial element of the Global Ocean Conveyor.

Here's how the Antarctic component of the conveyor works: As the south Atlantic portion of the conveyor moves into the Southern Ocean, the Coriolis effect diverts it to the east. Since this open ocean completely circles the globe at this latitude, there is no landmass to give the current a purchase to move farther south, except for very deep water that follows the contours of the sea floor. During the sea ice season, some of this already cool water freezes into sea ice, which has two effects. On the one hand, as it freezes, it releases a small amount of heat, and, secondly, as fresh water is captured in sea ice, the remaining seawater becomes saltier. This cooler, salty water is denser than surrounding waters and sinks to become what is called deep water.

The formation of what is called deep water is one of the main engines of this gigantic system of heat distribution in the oceans. As it sinks, this deep water pulls the conveyor as water moves in to replace the sinking water. There are only a few places on earth where such deep water forms. Two are off the coast of Antarctica, and the other two are in the far north, one spot between Iceland and Greenland, and the other in the Norwegian Sea. The way deep water is formed and sinks in Antarctica is thus slightly different than deep water formation in the North Atlantic, although the engine of both systems is the same.

Another major building block of the present form of the ocean conveyor and modern climate was the uplift of the Isthmus of Panama. This event entirely closed the ancient Central American Seaway between 2.5 million and 3 million years ago, blocking the westerly flow of water from the Atlantic to the Pacific. In turn, this set in motion a cascade of repercussions, creating the modern Gulf Stream as this new barrier deflected warm waters northward, while perhaps contributing to the glaciation of the north polar latitudes. Contemporaneous with these events was the onset of the most recent series of ice ages.

As the newly formed Gulf Stream moved northward, it warmed the climate of the North Atlantic. The main northern extension of the Gulf Stream is called the North Atlantic Current, and farther north this current becomes the North Atlantic Drift, a slow-moving wide expanse of water that significantly warms its surrounding area. Some estimates suggest that the heat released by the drift adds between 15 and 25 percent (depending on the latitude) to the energy northern Europe receives directly from the sun.

As the drift cools and as evaporation makes it saltier, the water becomes denser and heavier. In "normal" times, a good deal of the drift passes over a series of underwater sills, such as the underwater ridge that lies between Iceland and the Faroe Islands. Once over this sill, which averages several hundred meters in depth, the weight of the dense, cool, salty water causes it to plunge into the depths of the Norwegian Sea. Other lobes of this current sink in the Labrador Sea, and this cold, salty deep water then begins its journey south. As in the Antarctic, the "pull" of this sinking water is the engine that brings warm water north. Because the water entering the North Atlantic is so much warmer than that entering the Antarctic, however, it has a bigger impact on surrounding climate than the meager heat given up during Antarctic deepwater formation.

While most oceanographers stipulate the contribution of thermohaline circulation to Europe's climate, some oceanographers take issue with the degree to which this heat warms the continent. For instance, Richard Seager of Lamont-Doherty led a study published in the October 2002 issue of the *Quarterly Journal of the Royal Meteorological Society*. Based on observational data and climate models, Seager and his collaborators argue that almost all of the relative warmth of European winters can be explained by atmospheric transport of heat thanks to the disruption of the path of planetary winds by the Rocky Mountains (the mountains temporarily deflect the jet stream southward and the winds pick up heat and moisture released by the ocean in winter before they get to Europe), and by the release of ocean heat stored locally rather than transported by the North Atlantic Drift. In other words, Europe is warm because it lies on the edge of a continent on the eastern side of an ocean.

That being said, the authors of the study acknowledge that the THC does have important impacts above 60 degrees north, and that by secondary effects, such as dampening the formation of sea ice, it might impact weather farther south as well. Even the most vociferous proponents of the role of THC in Europe's climate do not ascribe all of Europe's clement climate to heat transported northward. Terrence Joyce of the Woods Hole Oceanographic Institution fixes the amount as between 15 and 25 percent, depending on the latitude, but notes that even at the lower end of that estimate the marginal warmth provides the difference between harsh frigid winters and the more manageable temperatures of more normal conditions.

There are many unanswered questions about the THC. No one is really sure why this deepwater formation only occurs in the Atlantic Ocean in the northern hemisphere. The North Pacific is less salty than the North Atlantic, which means that the ocean currents do not have the density to sink as they do in the Atlantic. Some climate modelers speculate that the rivers running back to the Pacific from the coastal mountain ranges of North America collect and return evaporation from the Pacific, continually pouring fresh water into the ocean. It's possibly that the bathymetry of the North Atlantic, with its underwater ridges and choke points, plays a role as well. Reconstructions of past movements of the North Atlantic Drift suggest that during the ice ages thermohaline circulation varied in intensity. At times it shut down or diminished, and during these cool periods, the North Atlantic Drift shifted to the south. These shifts played out repeatedly during the wild climate swings of the most recent ice ages, which began some 2.5 million years ago.

What caused the ice ages also remains an open question. Some experts on paleoclimate attribute the onset to the repercussions of the rising of the Isthmus of Panama (thereby diverting an enormous amount of water vapor northward, which provided the raw materials for ice sheets), while Mark Cane, an oceanographer based at Lamont-Doherty Earth Observatory (and one of Seager's coauthors), argues that uplift in the western Pacific laid the groundwork for these latest ice ages. Between 3 million and 4 million years ago, the northward drift of New Guinea and the surfacing of parts of the Indonesian archipelago would have rejiggered the ocean currents of the Pacific, replacing the flow of warm water from the South Pacific into the Indian Ocean with cooler waters from the North Pacific, while also changing the currents that formerly carried warm water into the upper latitudes. This latter change could have precipitated the ice ages, argue Cane and his MIT-based collaborator Peter Molnar, while the cooler waters entering the Indian Ocean would have reduced the rainfall for East Africa.

Whatever combination of events started the ice ages, the general cooling and drying was punctuated with great swings of climate, periods during which climate would remain unstable for 100,000 years or more. Ice ages tend to be more tumultuous than interglacial periods in no small measure because of the impact of ice on the ocean conveyor. But the fault lies in the stars as well, or at least in earth's orientation as it orbits the sun. Most likely, plate tectonics set the table for the beginnings of the ice ages, and earth's orbital dynamics supplied the chill.

The longest orbital pulse that changes climate involves the regular rounding and flattening of earth's orbit as it circles the sun. One full cycle takes about 100,000 years. This seems to coincide with the spacing of ice ages. As earth slowly chills at about .01 degree centigrade per century, ice builds up over a

90,000-year period. Then in the next 10,000 years, it all melts. Ice sheets have complicated and sometimes counterintuitive dynamics of their own. In *The Two-Mile Time Machine*, for instance, Richard Alley argues that big ice sheets can melt more rapidly than small ones, and that small changes in sunlight over the long term can have large impacts on ice sheets, while big changes over the short term have little impact.

Clearly, it's not as simple as "farther away during elliptical part of the orbit equals ice age" because where earth is during this orbit must also be understood in terms of whether an effect on climate is enhanced or muted by other aspects of orbital dynamics. For instance, another cycle involves the tilt of the planet as it travels through its orbit. During this journey, the angle of its spin axis gradually shifts between 22 degrees and 24 degrees and back. The inclination of this axis explains why the earth has seasons (one hemisphere will have nearly maximum tilt toward the sun at the start of the planet's annual orbit and then maximum tilt away 180 days later). Also, the degree of inclination toward the sun either exaggerates or diminishes the contrast of the seasons.

This is easy to envisage because the less tilt to the earth's axis, the less difference there will be in the amount of sunlight received by either pole at a given time. The greater the tilt, the more contrast between summer and winter temperatures. This constant seesaw between 22 and 24 degrees of inclination has a period of 41,000 years.

Then there is the wobble of earth's spin axis. Imagine the spin axis as a very long stick that entered the earth at the North Pole, went through the center of the planet, and exited the South Pole. The angle of that stick would vary between 22 and 24 degrees as just described, but the end of each stick would also trace a rough circle as the spin axis wobbled. This circle is called precession, and the wobble completes a circuit roughly every 19,000 to 23,000 years (the true period of precession is somewhat longer, about 26,000 years, but it is effectively shortened by the rotation of earth's orbit). Precession's effect on climate is to change which hemisphere has winter or summer when the earth is closest or farthest from the sun.

Precession becomes more or less significant depending on where we are in earth's 100,000-year oscillation between a flat and round orbit around the sun. The effect of precession is to change the angle of the rotation axis of the earth relative to the sun as it traces its 20,000-year circle, changing which hemisphere is tilted toward the sun at different points in earth's orbit. When earth is in the rounder part of its 100,000-year pulse between round and elliptical, precession is not so important, because distance from the sun would not vary much, only about 1.6 percent at present, versus 5 percent at the most elliptical. At those latter times, however, when earth's orbit tends more toward the squashed-circle shape, precession can change which hemisphere has greater contrast between summer and winter.

Imagine, for instance, the situation when earth's orbit is more squashed so that its distance from the sun varies more significantly over the year. If during this phase the precession of earth's axis is such that the northern hemisphere is tilted away from the sun when the earth is farthest from the sun in its orbit, and tilted toward the sun when earth is closest, then the net effect would be to exaggerate the difference between summer and winter in the northern hemisphere. In the southern hemisphere, however, the situation would be the opposite since it would be summer when earth was receiving less radiation and winter when it was receiving more, reducing the contrast between the seasons.

These big cycles have an impact on the occurrence of ice ages because they change how much solar radiation hits the earth, where it hits and when. The recipe for growing ice sheets, for instance, depends more on short, cool summers than it does on how cold it is in the winter. Key variables are how long temperatures are below freezing and how little they rise above it, not how far below freezing temperatures plunge.

The long climate cycles are like an ultralow-frequency ringing, and like any note when struck, they have harmonics. Ice core specialist Paul Mayewski of the University of Maine speculates that another 11,000-year cycle revealed by the ice core data represents just such a "harmonic" or ringing set in motion by the reverberations of the longest cycles.

Over the years, specialists in paleoclimate have noticed several other cycles out of ice cores taken from ice sheets and glaciers, from sea- and lake-bed sediments, from tree rings and other encrypted records of the past. For instance, during the recent ice ages, there was a regular cycle that saw temperatures fall and then rise over a roughly 6,100-year period. Described first by Hartmut Heinrich, a pioneering German marine geologist, these events occur during the coldest part of an ice age. According to Mayewski, Heinrich, Gerard Bond of Lamont-Doherty, and other climate scientists, this cycle is likely a lagging response of ice sheets to changes in solar radiation.

As the ice sheets grow, their internal dynamics eventually cause them to spread out into the seas and eventually peel off armadas of icebergs. Richard Alley argues that this is because, as ice sheets grow, they trap more and more heat underneath them, eventually melting their frozen bond with the underlying earth. This melted layer allows the sheet to surge outward. In turn, the flood of icebergs would cool the ocean and air. That would lead to an increase in sea ice, which would further cool the air by reflecting sunlight and trapping ocean heat beneath the ice.

Many scientists see the sea ice as a necessary amplifier of abrupt climate shifts. Jeff Severinghaus of the Scripps Institution of Oceanography describes one way this might happen. A shutdown of the conveyor sets the stage for a rapid increase in sea ice (in Antarctica, during winter, sea ice expands by 30,000 square miles a day), and once sea ice covers the northern oceans, "the atmosphere thinks it's a continent," he explains. Continents have no memory for temperatures. What this means is that they can't retain and dole out heat the way an ocean can, except on very short timescales. Moreover, the white surface of ice is about eight times more reflective of heat than open water, so the spread of sea ice provides a natural amplifier of cooling by both trapping the relative warmth of the ocean beneath it and by increasing the amount of heat reflected back into space.

The transfer of so much ice to the oceans sets in motion other cycles. Alley suggests that diminished ice in Hudson Bay, for instance, would reduce the cooling influence of winds flowing off the ice sheets. Moreover, all this fresh water introduced by icebergs would at first shut down thermohaline circulation and then, perhaps as more and more fresh water was taken up and sequestered in sea ice, eventually restart the conveyor on a roughly 1,500-year cycle. Noteworthy about these Dansgaard-Oeschger cycles (as they are named) is that the cooling and warming at either end occurs quite suddenly, with major warming and cooling shifts taking place in as little as a few years. The late Gerard Bond of Lamont-Doherty identified a pattern, now called a Bond cycle, in which these events become progressively cooler following each big Heinrich event, until the cycle ends with a very big warming.

Complicating all of this are both the big orbital cycles as well as more frequent cycles of solar variability caused by events such as sunspots. The effects of volcanic eruptions and the occasional comet striking the planet also skew the climate record. No wonder Richard Alley was moved to use the following metaphor for climate cycles, which sounds like something out of Gilbert and Sullivan: "You might think of a roller coaster riding the orbital rails, with Heinrich-Bond jumping off the roller coaster while playing with a Dansgaard-Oeschger yo-yo."

In the course of the past 100,000 years, there have been seven Heinrich events (they were marked by bigger intervals at the beginning because earth's orbital position was less conducive to growing ice sheets). The last one—dubbed Zero—was the Younger Dryas; it was, as noted, a doozy, plunging temperatures by as much as 27 degrees Fahrenheit in parts of the world in less than a decade. Beginning roughly 12,700 years ago, it ended about 1,300 years later with an equally dramatic warming that led into the current warm period.

Eclipsed by seismic climatic events such as the Younger Dryas are smaller perturbations of climate that take place over much shorter timescales. The North Atlantic Oscillation and the Pacific Decadal Oscillation describe regime shifts in atmospheric pressure over the North Atlantic and North Pacific that seem to change over twenty-year periods. The now familiar El Niño reworks storm tracks and rainfall patterns around the world every few years. When compared with the major climate shifts of Dansgaard-Oeschger cycles or Heinrich events, even the most severe El Niño would barely budge the needle. Even so, the 1997–98 El Niño did more than $100 billion damage to the global economy and killed tens of thousands of people.

There are other, more subtle influences on climate than changes in Earth's orbit, its attitude toward the sun, the oceans and ice, and the various harmonics of these systems. One is the composition of the atmosphere. If earth gets a fixed allowance of heat from the sun (at least on human timescales), the amount of the allowance that is retained has a great deal to do with the composition of the atmosphere. Incoming solar radiation is either reflected, absorbed by the earth, or sent back out toward space as infrared radiation. How much of that infrared escapes depends on the properties of the atmosphere. There are fourteen gases in the atmosphere, and most of them have little ability to trap outgoing infrared radiation. Those that do, such as CO_2, methane, or nitrous oxide, are called greenhouse gases because, like a greenhouse, they keep heat from escaping back to space. Methane traps a lot of heat, but it is much rarer than carbon dioxide. It doesn't take a lot of a greenhouse gas to affect temperatures.

Atmospheric levels of carbon dioxide, by now the most familiar greenhouse gas, have risen or lowered in step with global temperatures for millions of years, although scientists only have reliable ice core data for the past 400,000 years. For instance, if there were 250 parts per million of CO_2, earth's average temperature would level at about 57 degrees Fahrenheit. Now with 380 parts per million (the highest levels in 500,000 years), temperature has risen about a degree. If we reach 880 to 1,000 parts per million as expected within the next 100 years without reductions in emissions of carbon dioxide, the atmosphere will have heat-trapping abilities not seen for at least 30-to-40 million years, according to Daniel Schrag of Harvard, and at that time the planet was astoundingly hotter than it is now. Just a hundred or so ppm more than those levels and the atmosphere would resemble that of the Cretaceous period, when dinosaurs roamed on an earth described by Richard Alley as a "saurian steambath." In 2004, Schrag and Richard Alley published an article in *Science* in which they noted the disturbing fact that no present-day global climate model can produce the temperatures evident in the paleoclimate record from that time, raising the possibility of presently unknown feedbacks that might amplify a runaway warming.

Whether changes in CO_2 lead or lag changes in temperature, and why levels rise and fall remains a matter of debate, but the correlations are strong. Writing in *Science*, Richard Kerr quotes Thomas Crowley of Duke University: "You can't say CO_2 explains everything, but it does explain a heck of a lot." It is worth stressing that there is no dispute about the physics through which CO_2 and other greenhouse gases retain heat in the lower atmosphere.

The most important greenhouse gas is water vapor, and it's also the most elusive. It's exceedingly difficult to reconstruct past cloud conditions. Clouds both reflect and trap heat, and whether they warm rather than cool the lower troposphere depends on how thick and how high they are. In the week following the terrorist attacks of September 11, 2001, the Federal Aviation Administration shut down air traffic, which meant that for at least that week there were no cloud contrails in the upper atmosphere. Not coincidentally, the contrast between day and night temperatures increased that week.

Any account of the orbital and other factors affecting climate naturally prompts the question "Where we are now?" Given all these moving and interacting parts, is it possible to pin down where we are now, or should be, according to the alignment of the earth, the dynamics of ice sheets, and other factors? The answer is yes. Sort of.

In terms of orientation in space, the earth is presently in the rounder part of the 100,000-year orbital pulse, which places us at the beginning of the 100,000-year ice age cycle. The tilt of earth's spin axis is inclined about 23.5 degrees, which by itself would accentuate the difference in seasons, but this effect is offset because of earth's current position in the precession of its axis. Presently, the northern hemisphere is tilted away from the sun when the earth is closest to the sun in its orbit, and toward the sun when it is farthest. Thus, during the northern winter when light hits the north less directly, that hemisphere is getting about 3 percent more radiation by virtue of earth being slightly closer to the sun, reducing summer-winter differences. At the same time, these seasonal differences are slightly exaggerated in the southern hemisphere, but here the effects of precession are tempered by the predominance of ocean water, which moderates seasonal shifts.

Given all the different factors that can screw up climate, we can thank our lucky stars for the rare syzygy of offsets that has been in place for 10,000 years or so. This is about as good as it gets in terms of orbital alignment. In *The Two-Mile Time Machine,* Richard Alley places us in the sweet spot of recent climate cycles.

Alley notes that the most stable periods in these 100,000-year cycles are the 10,000-year spans of coldest and warmest weather. We're now in that nice warm phase, and we would have to go back about 115,000 years to the Eemian interglacial era to find an equally warm and stable period. To extend Alley's metaphor of the climate roller coaster, these geologically brief warm periods are the equivalent of the pause at the top of the first big climb before the car begins its dizzying series of twists and turns and dips. If future climate carried forward what has happened in the past (and if humans had not evolved to rewrite the script), sometime in the future we would begin the next 90,000-year period of cooling, which would be an epoch marked by growing ice sheets and punctuated every few thousand years by violent, rapid, and extreme shifts as the ice grew. Based on the study of a European ice core project from Antarctica that provides a 740,000-year record, however, some paleoclimatologists argue that present conditions are more like the situation of about 400,000 years ago when a warm period lasted for over 25,000 years. If this turns out to be the case, we've got some time before we begin the plunge, unless we screw things up through our alterations of the atmosphere.

The Holocene may be a protective bubble of warmth among the ice ages, but ripples of climate change have occasionally intruded to upset the calm. Like *mementi mori,* abrupt climate events have periodically interrupted this halcyon period, reminding humanity that it thrives in a bubble. The instant chill of 8,200 years ago arrived after a warm stretch of a few thousand years. Big as this event was, however, Richard Alley estimates that it was only half as strong as the Younger Dryas.

Then, with the planet aligned so that conditions were favorable for hot summers and cold winters in the northern hemisphere, the climate entered a warm period that lasted until about 5200 B.P. (or 3200 B.C.). Warming early in this interregnum raised sea levels sufficiently to allow Mediterranean water to breach the Bosphorus and flow into the Black Sea, then about 500 feet below sea level. Although some paleoclimatologists challenge this idea, Lamont-Doherty marine geologists Walter Pitman and William Ryan argue in *Noah's Flood* that the breaching of the Bosphorus created an enormous cataract, two hundred times as powerful as Niagara Falls.* The monster waterfall raised the level of the Black Sea by 6 inches a day, drowning fields and trapping the unwary. According to this theory, refugees from the rising waters scattered both east and west, bringing agriculture to Europe. Moreover, Pitman and Ryan argue, ancestral memories of this unforgettable flood later surfaced in the book of Genesis and *The Epic of Gilgamesh.*

Apart from prompting biblical floods, the mild weather provided a benign context for the first civilizations, notably in China, Persia, and India about 6,000 years ago, and then 1,000 years later in Egypt. Human population, which UNESCO estimates grew from 5 million to 7 million between

*William Ryan and Walter Pitman. *Noah's Flood: The New Scientific Discoveries About the Event That Changed History* (Simon & Schuster, 1999).

10,000 B.P. and 6000 B.P., more than tripled in the next 2,000 years, as innovations such as irrigation increased agricultural production and provided some protection against the vagaries of the weather. By the dawn of the Christian era 2,000 years later, the population had grown by a factor of ten. During the next 2,000 years, human numbers grew by a factor of twenty-four. Now, every two weeks, we add numbers equivalent to the human population of the globe at the beginning of the present warm period. Clearly the Holocene has been a good time, at least for us.

There has been no event since the 8200 B.P. episode anywhere near the magnitude of that sudden freeze. Still, even small cyclical changes in climate (compared to the epic swings of climate during ice ages) can have outsized effects, particularly when a society has prospered around growing a particular crop in a particular place. There have been many such midsized events. Paul Mayewski's interpretation of ice cores taken from GISP2 in Greenland is that there were rapid climate-change events between 6,100 and 5,000 years ago, between 3,100 and 2,400 years ago, and one that began 600 years ago known to all as the Little Ice Age.

10

Proxy Wars I: Ice

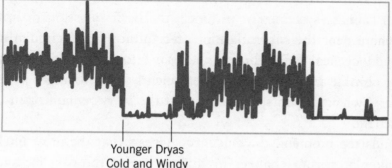

Younger Dryas
Cold and Windy
11,500 B.P.—12,700 B.P.

DUST IN ICE REVEALS DRASTIC CHANGE

VIEWING CLIMATE in terms of orbital cycles and other harmonics tells us much about where climate is heading on big timescales, but knowing where we are poised on Richard Alley's climate arc still leaves many questions. Paleoclimatologists want to know whether there have been abrupt climate events during the Holocene, for instance, and if so, what caused them? Do the rules that apply during glacial times apply during warm interludes? The relevance of these questions should be obvious: if climate changed abruptly during historical times, it can do so again. If the rules of glacial-era abrupt changes still apply, then we can better understand the scope and intensity of past events as well as better understand the signals that indicate another abrupt event might be upon us.

The MIT theoretician of artificial intelligence Herbert Simon wrote that solving a problem is nothing more than representing the problem in such a way that the solution becomes obvious. In the case of climate, a clearly stated question—do the rules of abrupt climate change apply in warm periods as well as glacial?—does not necessarily point to a clear way of representing the problem so that the solution becomes obvious. For one thing, large gaps remain in our understanding of the climate system. For instance, the basic question of how changes in ocean circulation affect atmospheric circulation and vice versa leads climate scientists into that hall of mirrors of bewildering feedbacks. When scientists put together a scenario of past events, large parts of the story remain open to debate.

At the moment, the evidence suggests that the most likely cause of abrupt changes in climate during the past ice ages was a sequence of events in which thermohaline circulation shut down, fostering the conditions for an increase in sea ice, which in turn greatly amplified and broadcast the cooling far beyond the region of the North Atlantic. It is for this reason that paleoclimatologists are interested in whether this circulation has ever shut down during the Holocene, and if it has, how such a shutdown differed from the catastrophic changes brought about by shutdowns during the last ice age.

Paleoclimatologists remain divided on whether THC has shut down since the last ice age ended, as well as how a partial or complete shutdown, if it happened, would have changed climate. Some records show abrupt changes, and proxy records suggest past shutdowns of the ocean conveyor, but some of these records are open to differing interpretations, and with each year, as physical oceanographers learn more about the complexities of the circulation, they become less

confident that there is a simple and direct path between melting episodes, shutdown of THC, and freezing temperatures in
Europe. One indicator of the rapid pace of discovery has been
that the name of this critical ocean circulation has changed
twice since Broecker proposed a "Global Ocean Conveyor" in
the late 1980s, first to thermohaline circulation, and now to
meridional overturning circulation, or MOC. The reason for
the most recent change is that THC contains the assumption
that changes in salinity drives the conveyor, while MOC
simply describes what is happening, leaving open what is driving the system (many scientists, however, stick to THC for
convenience' sake). This steep learning curve is a product of
an ever richer portrait of Holocene climate as proxy records
pour in from the far corners of the globe. Since this record
drives both the theory of climate change and provides the
basis for assertions of climate's role in historical events, it's
worth a deeper, more detailed look at the climate forensics
that link chemical traces to global statements about temperature, wind, and rainfall. If climate is to be tagged as a serial
killer of civilizations, the question must be asked: how credi-
ble is the witness, namely the proxy record?

The answer will always be relative, since the perfect proxy
does not exist. Ask ice core specialists about speleothems (the
climate record contained in cave formations), and they will acknowledge that the stable conditions in caves provide an attractive place to look for a record of the past, but the same
scientist will mention that rainfall differences can distort the
record. Ask ocean-sediment specialists about tree rings, and
they will acknowledge that tree rings provide a peerless record
of annual weather changes in a particular place, except for the
caveat that the record might also be distorted by rainfall differences for trees in the more variable mid latitudes.

And thus it goes—ocean sediments provide a good long-
term record, but just a tiny piece of the ocean has been drilled

and a minor shift in the location of ocean currents might leave a record that looks for all the world like a momentous and abrupt climate shift. Ice cores provide a great long-term record, but the cores coming from the bottom of a glacier or ice sheet might be distorted by folding and other internal changes in the ice sheet that are invisible from the surface. Tim Barnett of the Scripps Institution of Oceanography points out that none of the ice proxies or coral proxies match the observed record for the past hundred years.

Most paleoclimatologists will readily admit the limitations of their own proxy. All will stipulate that there is no perfect record of the past. Perhaps the best way to convey the strengths and limitations of various proxies is to go through the natural process by which these records of past climate become encoded in ice, sediments, and living material, and then to rewind the chain of logic by which different scientists identify and unlock these records.

Let's take ice, for instance. Ice forms toward the top of an ice sheet as each year's accumulation of snow gradually compresses the snow beneath it. As the layers accumulate, the snow first becomes *firn*, a German word used to describe the friable intermediate state of snow before it becomes ice. Entombed with the ice are air, acids produced by interactions of atmospheric chemicals with sunlight or water, dust blown in, sea spray carried by the winds, and volcanic ash. While dust, ash, and sea spray mostly stay in place in the order in which they fall on the snow, it can take a couple of hundred years for the snow to become sufficiently compressed for gases to become isolated in bubbles. As the accumulating ice continues to weigh ever more heavily on the ice beneath it, the bubbles eventually disappear as the gases are forced into spaces between the frozen water molecules. They only reemerge when the ice is released from the tremendous pressures of an ice sheet.

Ice coring as a means of understanding climate history is a relatively new field in America, dating back to the International Geophysical Year, 1957, when scientists from around the world mounted a serious effort to take earth's pulse. Since then, scientists have been learning on the fly. Apart from developing reliable methods to decode proxies, they face the difficulties of handling ice in the harshest environments on earth.

Ice cores have to be treated like gelignite as they are brought to the surface. A piece of ice simply pulled up from the interior of the sheet can easily explode if it is simultaneously warmed and released from pressures 154 times greater than at the surface. At first the ice extracted from deep in an ice sheet is clear because the pressures so compress the air bubbles. To prevent these cores from getting the bends, the ice spends a year in chambers chilled to –20 degrees centigrade, where it can "relax." Over the year, some of the ice turns cloudy. Then various geochemists and physical oceanographers have a go at the cores. The GISP2 scientists, for instance, took fifty measurements for each section. For the first 11,000 years, they sampled every 2.5 years; from 11,000 to 40,000 years ago, they sampled every 15, and from 40, 000 to 110,000 years ago, every 50 years. The ice record goes back further—at least 250,000 years, perhaps as much as 400,000 years—but toward the bottom of the ice sheet, folding and deformation takes place as the ice spreads out and confuses the record. At the top of an ice sheet, a year's worth of data might be contained in a 1-meter-long core, while at the bottom a meter might contain a millennium of compressed accumulation.

To calibrate the tape of history contained in an ice core, geochemists and other researchers want to know exactly how the accumulating ice makes a record of seasonal changes and other events. First, that means looking at the layers. To do that, they construct what is, in essence, a light box in the top of the ice sheet itself.

I visited one such apparatus a few years ago on the West Antarctic Ice Sheet. It consisted of two snow pits, each more than 7 feet deep and separated by a wall of intact snow and ice about 1 foot thick. One pit was covered with plywood so that the only light entering that side came in filtered through the 1-foot-thick wall. Researchers stand in the covered pit and look at the translucent wall to see a visual record of accumulation. The light filtering through the snow leaves an impression similar to the decorated rice paper typical of traditional Japanese walls. Symmetrical lines marking seasonal layers create an elegant design.

Christopher Schuman, a geochemist who has worked in both Greenland and Antarctica, walked me through this particular tale from the crypt. I was standing on the layer laid down during 1990–91, and in the snow and firn wall separating the two pits I could clearly see the thin layer from 1994–95 when there was very little accumulation. The boundaries marking the layers are a product of the summer sun acting on the ice surface. While the temperatures almost always remain well below freezing, the summer sun hitting the snow causes what's called sublimation as snow becomes water vapor without melting. During cooler parts of the day, this moisture migrates and recrystallizes, forming a frost that has a distinctly different character than winter snow.

The visual record is just the beginning. At the time of my visit to Antarctica, Schuman was calibrating the ratio of oxygen isotopes to develop the temperature proxy. Then he would compare the peaks and valleys with satellite records of microwave radiation from the same period to develop a profile of the timing of snow accumulation, which, he said, was critical to interpreting ice formed in prior years.

Understanding why requires a little background on why scientists were drawn to isotopes in the first place. An isotope of an element is an atom that has an extra neutron or two. Oxygen, for instance, usually has eight protons and eight neutrons, giving it an atomic weight of 16, but variants of oxygen occur naturally with nine or ten neutrons, so that an examination of an array of oxygen atoms will also find O_{17} and O_{18} represented. In the 1960s, the Dutch geochemist Willi Dansgaard figured out that the ratio of heavy oxygen to lighter oxygen atoms in ice might be used to determine ancient temperatures. Cool temperatures tend to wring moisture out of the air, and because water molecules containing O_{18} tend to be heavier, they condense first. Thus, as cold intensifies and continues, the oxygen atoms in precipitation tend to get lighter and lighter since most of the heavier O_{18} has already fallen out of the water vapor in the air. By capturing old water from ice and measuring these ratios, a clever geochemist might compare those ratios with the known ratios characteristic of different temperatures today and thus reconstruct a thermometer for past climate—assuming of course that those ratios have remained constant over time. One of the remarkable scientific achievements of the past fifteen years has been the development of fiendishly clever ways of calibrating these ice-embedded proxies, and indeed, many of these breakthroughs have taken place in the years following the first publication of the GISP2 results from Greenland.

GISP2 is but one of several ice core projects of the past fifteen years. There are two major drilling projects at various stages of completion in Antarctica, and Paul Mayewski, Lonnie Thompson, and others have drilled glaciers from the Himalayas to the Andes for detailed records of more recent climate changes. As many tropical glaciers retreat around the world, some of these measurements come from ice that has already melted (high-altitude equatorial climates are particularly stable so that the melting of tropical glaciers offers dramatic evidence that global climate is changing). The startling findings of GISP2, however, created the momentum that facilitated both new drilling projects as well as new techniques of analyzing the data encoded in ancient ice.

It takes a special type of scientist to have the ruggedness, analytic tools, engineering savvy, and patience to study ice cores. One such is the glaciologist Paul Mayewski of the University of Maine. He cut his teeth in Antarctica and was chosen to head the scientific team for GISP2. Beginning in the summer of 1989, it took the team five summers to extract the full two-mile core.

By the middle of June 1992, the team had extracted cores from over 1,500 meters deep in the ice sheet. At this depth, the ice had first been deposited 4,200 years ago. Then, the next summer, they pulled up ice from the years between 10,000 and 15,000 B.P. During the analysis of these cores, Schuman, who worked on GISP2 before his work in ice pits in Antarctica, helped out, filling in for a colleague who had hurt his back. Looking at the stratigraphy, the visual record of events seen in a vertical slice of the core, for ice from about 11,000 years ago, he saw a sharp line. Alley said, "You've got it," meaning that Schuman was looking at the particular events that overturned the understanding of how climate changed from warm to cold.

As Alley recalls, he and others felt a palpable sense that they were changing the understanding of climate when they first examined that ice core. One of Alley's assistants, Wanda Kapsner, pointed him toward the smaller crystals in the ice core that signaled the Younger Dryas transition. They then sliced the core along its length, preserving the stratigraphy. Kendrick Taylor, an ice core specialist based at the University of Nevada's Desert Research Institute, took electrical measurements that showed an abrupt change in electrical conductivity at that boundary, a confirming signal of rapid change. Then Alley began looking at the chemistry of the ice.

A few months later, Alley, Taylor, and colleagues published a paper in *Nature* entitled "Abrupt Increase in Snow Accumulation at the End of the Younger Dryas Event." Dansgaard had already proposed the idea, but this paper had far more impact because, with nearly a mile of ice beneath the level this was taken, there was little possibility that the changes were produced by the response to the ice of uneven bedrock underneath. Instead of thinking about climate as a dial that could be turned up or down, as Alley has analogized it, the scientists were proposing that climate controls were more like a switch.

This hypothesis was as explosive as deep ice itself. Around the world, scientists redoubled their efforts to refine the precision of interpretation of the proxy record that pointed toward such rapid changes. For instance, Chris Shuman knew that the nice chain of inference that linked oxygen ratios to changes in temperature could be thrown off if the snow that fell on the ice sheet shifted from falling mostly in winter to falling mostly in summer. Summer snowfalls would tend to have more heavy oxygen isotopes; thus, even if the temperature cooled, the record might suggest warming if all the snowfall simply shifted from winter to summer. It was to eliminate this ambiguity that Shuman devoted his efforts to determining not just how much snow fell during a given year, but when it fell.

Other geochemists have worked on other ways to calibrate the isotope paleothermometer. Up in Greenland, notes Richard Alley, the ice scientists used temperature readings taken from the borehole itself to help calibrate their oxygen isotope thermometer. According to Alley, deep in the ice sheet, the ice retained some memory of past temperatures, and the ice sheet's memory might be tapped by dropping a thermometer down the borehole and taking readings for various depths. The borehole temperatures provided a low resolution record that, in Alley's words, tends to "forget short-lived events" by smoothing over abrupt transitions, but they could be used to "ground truth" the readings implied by the isotope ratios.

To get a proper picture of past climate, the researchers not only needed a record of past temperatures, but also an indication of how fast those temperatures changed. Because gases could move around inside the snow for more than a century before they became sealed in bubbles, scientists needed a way to determine which isotopes gave a reliable measure of temperature change on shorter timescales. Jeffrey Severinghaus solved that problem by developing two additional gas fractionalization tests that would indicate abrupt changes in temperature.

This solution too grew out of knowledge gained from snow pits, specifically how gases move within the top of an ice sheet before they are imprisoned in bubbles. In the two hundred years during which the weight of compressing snow turns firn into ice, the gases separate in different ways than they do in the atmosphere, where convection and air currents stir everything up. By contrast, in the snow, gases tend to move because of gravity, and because of temperature, with the heavier isotopes tending to migrate toward colder regions when temperatures were changing. By measuring this thermal diffusion with an understanding of aeration in snow and controlling for the other factors affecting the movement of gases, Severinghaus reasoned that he could come up with a way to reduce the uncertainty about events during the years before the gases become trapped in bubbles. Specifically, he focused on the ratio of isotopes of nitrogen and argon gases trapped in air bubbles.

He chose this ratio because both isotopes settle at the same rate because of gravity, but the argon isotopes are only 60 percent as sensitive to temperature changes as nitrogen. Thus, the relative enrichment of nitrogen isotopes trapped in air bubbles provides a signal of temperature change. Moreover, Severinghaus showed that the signal provides a precise gauge of when those changes occur. As he and his collaborator Edward Brook wrote in a 1999 article in *Science,* "After an abrupt climate warming, a temperature gradient will persist in the firn for several hundred years (the thermal equilibration time of the firn) and will thermally fractionate the entire air column. Gases diffuse about 10 times as fast as heat in polar firn, so the isotopic signal penetrates to the bottom of the firn long before the temperature equilibrates, in about a decade. The bubbles thus record a signal of the climate event as an abrupt increase in N_{15}, slightly smoothed by the diffusion process in the firn, followed by a gradual decrease in N_{15} over several hundred years as the firn becomes isothermal once again."

Richard Alley, Wallace Broecker, and a number of other paleoclimatologists described Severinghaus's test for ancient temperature change as a brilliant breakthrough. Indeed, Broecker describes Severinghaus as the best student he ever had. Refining the precision of temperature changes to within a decade, for instance, allowed Severinghaus to weigh in on one fundamental question of abrupt climate change: whether abrupt warmings begin in the tropics or in the northern regions.

Paleoclimatologists favor methane as a proxy for warming because in preindustrial times methane largely varied according to biological activity in wetlands—an indication of warming. Moreover, the gas gets thoroughly mixed in the atmosphere, and remains in the air for no longer than ten years. During glacial times, ice sheets covered the vast wetlands of the far north, so amounts of methane retrieved from ice age cores tended to reflect warming in more tropical regions. The O_{18} ratios obtained from ice core records showed that the last glacial maximum ended with an abrupt warming about 14,670 years ago. Severinghaus confirmed this figure using nitrogen and argon gas ratios, but the accuracy of this latter data allowed him to show that nitrogen ratios began to show temperature change a good twenty years before methane levels began to creep up, meaning that warming in the Arctic preceded the response in the tropics. Numerous other proxy records obtained from seabed sediments, pollen, and other sources confirm this finding. Because methane responds quite quickly to changes in temperature, Severinghaus argues that the time lag indicates that the abrupt warming started in the North Atlantic and then, some decades later, the tropics responded.

Unfortunately, the result poses a new headache for paleoclimatologists, many of whom naturally would expect the tropics to drive global climate because that's the area that receives the most energy from the sun. While many scientists can understand how changes in the North Atlantic, particularly a massive increase in sea ice, could cool the world, no one yet has an equally compelling argument about how the North Atlantic might kick-start rapid warming.

Which leads us to a sore question in climate science: whether Greenland provides a good proxy for weather elsewhere. "Greenland is a funny place," says paleoclimatologist Dan Schrag, chairman of Harvard University's environment program. "The center of the ice sheet is two miles high, and Greenland's climate is incredibly sensitive to changes in sea ice, so ice cores from Greenland are not necessarily the best proxy for reconstructing the climate of the northern hemisphere."

Schrag is an influential climate scientist, and his cautionary warning about Greenland must be taken seriously. In March 2000, he published an article in *Nature* entitled "Of Ice and Elephants." In it, he analogized the way different climate scientists look at the determining aspects of climate to the old story of the blind man and the elephant, in which one man, feeling the trunk, says it is like a snake, another, feeling the ear, says a leaf, and another, feeling the leg, says it's a tree. "People who study Greenland ice cores think the global climate revolves around their part of the world, as do those who study Antarctica, and those who focus on the tropics," he told me, "and you know what? They are all right!" Schrag went on to explain that while each place is important, "climate is so complicated that it's impossible to single out any one place that drives all others." Still, Schrag acknowledges that while establishing a direct chain of linkages between Greenland and western Asia might be difficult, both can be affected by changes in the jet stream.

The task of connecting events in Greenland to the rest of the climate "elephant" was not something that needed to be solved solely through the examination of ice cores. Even while geochemists were refining their ability to decode the encrypted messages in the ice, other scientists were finding evidence supporting the findings of GISP2 in other proxies collected far from Greenland.

11

Proxy Wars II: Mud

A Section of Deep Sea Sediment Core

THE STUDY OF ice cores from Greenland provided an extraordinary picture of climate over the past 100,000 years and
opened the eyes of paleoclimatologists to the possibility of
abrupt climate change. To use the analogy of a rocket heading
into orbit, GISP2 was the first stage in propelling the idea of
abrupt climate change into scientific consciousness. If the idea
was to achieve exit velocity and become established as a new
scientific paradigm, however, it needed the thrust of additional
proxy evidence, preferably not from ice, and definitely not
from Greenland. Fortunately for both the paradigm and the
field of paleoclimatology, the geologist Gerard Bond decided
to apply methods of analysis he developed studying rock out-
croppings to the analysis of seabed sediments. Moreover,
Bond produced the first of what was to become a veritable
mud slide of detailed evidence of past climate extracted from

seabed sediments around the world, evidence that has both deepened understanding of rapid climate change but also raised new questions.

Before he went to work studying deep-sea cores at Lamont-Doherty, Bond had spent some years looking at Cambrian rocks in western Canada, and had developed a way of using color patterns in rocks to interpret long-term climate cycles determined by orbital patterns (the climate cycles were called Milankovitch cycles in honor of the Serbian mathematician who first described them). At Lamont, Bond proposed looking at some well-studied deep-sea cores taken from the sea floor off of Ireland to see if his methods might help date the changes in those sediments. Because geologists sample differently than geochemists (who look at samples taken from equal intervals in the core), he thought that his fresh eyes and different approach might yield some new results even in a core that had been well studied.

As a matter of courtesy, he sent his proposal, along with the color patterning in the deep-sea core, to Wallace Broecker. At that point, Broecker was interested in Milankovitch cycles, but he had other fish to fry as well. He got excited when he saw color patterns corresponding to the 1,500-year abrupt climate-change cycle first proposed by Willi Dansgaard. At first, Bond was confused by what Broecker wanted. As he put it when I interviewed him (Bond died of cancer on June 29, 2005), "I'd never heard of Dansgaard-Oeschger events, and barely knew about ice cores." Still, at Broecker's urging, he sought permission to sample one particular core at 1-centimeter intervals to see what was causing color variations in the sediment. This was a much more fine-grained analysis than any geochemist had done at that time—the shortest time frame in a Milankovitch cycle is 20,000 years, so geochemists could get results by looking at 5,000-year intervals.

At first, noted Bond, people thought it was ridiculous that

he was focusing on something as simple as changes of color. He also had to negotiate the complicated and highly competitive politics of this new frontier of climate studies. For instance, he needed to examine cores extracted by the GRIP scientific consortium, a European ice-coring project that was drilling at the same time as GISP2, but to get their cooperation, he had to sign an agreement not to publish his results before they published theirs. He quickly won converts, however, when his approach got results. He discovered that the dark layers were from sand and gravel sediments transported by ice, while the lighter banks were largely composed of foraminifera (popcornlike creatures) that thrived in warmer temperatures. Bond, Broecker, and colleagues published a paper in 1993 in *Nature*, in which they established the correlations between the Greenland and North Atlantic sediment records. Then, a year later, Broecker published a paper in *Nature* that detailed the pattern of abrupt climate change events. Broecker named them Bond cycles in honor of Bond's groundbreaking work. Then, in 1997, Bond and a number of colleagues published their findings that Dansgaard-Oeschger events continued into the Holocene. This publication formally buried the notion that the present climate era has been monotonously stable.

Apart from helping to redefine the understanding of climate, one of the most noteworthy aspects of Bond's work was that he and his colleagues found their dramatic results in a deep-sea sediment core that had already been sampled and studied for many years. Different scientists often look for different things in the same core. As Gerard Bond discovered, physical oceanographers looking for long-term changes in core 609 taken from the Irish Sea completely missed the signal of rapid climate change in the same core. In geology as in life, one tends to see what one has been trained to look for.

The image of a climate that periodically goes haywire was

there all along, just waiting for someone to look at the core in a slightly different manner. To be fair, time scales vary wildly with different sediment cores. Peter deMenocal says that he looks at cores with 5-millimeter resolution to see long-term changes in sediment cores that might contain 100,000 years of data in a 4-foot segment. At Lamont-Doherty, he showed me one such core taken from the sea floor off Mauritania. On this core, the span of a hand covers 20,000 years, and the last ice age stands out with its bricklike color.

On the other hand, if he or other scientists want to look for evidence of abrupt events, they have to seek places with rapidly accumulating sediments in order to uncover for climate change signals that might be otherwise missed, and they have to scan the sediments at much finer resolution than every 5 millimeters. Perhaps the extreme example comes from Gerald Haug who produced an extremely detailed picture of climate in Mesoamerica during Mayan times. To obtain a climate record detailed enough to correlate with historical events in Mesoamerica, he passed lake-bed sediment cores under an X-ray fluorescence unit with a depth of 10 microns, because only resolution that fine can detect changes in seasonal laminations that were less than a millimeter thick. In the places where a great deal of sediment becomes laid down, the span of a hand might cover hundreds rather than thousands of years.

The detailed sediment picture tells you what happened, and to a degree it can help say why. In the years since Bond found corroborative evidence of the story of abrupt change told by the ice cores, various scientists have sought to tackle lingering ambiguities inherent in the evidence produced by examining foraminifera. Some scientists wonder whether other factors such as nutrients in the seawater or simple meanderings of ocean currents might affect the picture produced by foraminifera.

Jerry McManus of the Woods Hole Oceanographic Institution (WHOI), for instance, tries to find proxies that are immune to distortion by biological and other nonclimate changes. McManus acknowledges that carbon isotopes taken from forams show big changes in temperature at different times in the Holocene but worries that different factors such as changes in sea ice can skew that signal. So he uses multiple proxies. One of the most intriguing is designed to get around the potentially confusing signals coming from carbon isotopes, namely the relationship of two daughter atoms that result from the decay of uranium in the ocean.

McManus's proxy is based upon the ubiquity of uranium in the ocean. The uranium atoms he is concerned with—U-238 and U-234—both decay at very slow rates. Because they are relatively stable for hundreds of thousands of years, the atoms become well mixed throughout the oceans. As they decay, the two types of uranium produce two daughters, thorium and protactinium, and they produce these elements at a known rate and in a known relationship to each other. Neither "daughter" atom readily dissolves in seawater, so both should stick to particles in the water column and fall to the bottom at the same rate. What caught McManus's attention is that thorium is roughly ten times stickier than protactinium. What this means is that thorium tends to settle into sediments in twenty to thirty years, while protactinium takes over two hundred years to fall to the bottom. In that difference, McManus discovered an opportunity to determine when deepwater circulation shut down in the past.

Here's how it works: Since the Atlantic water cycles every two to three hundred years, the thorium tends to settle fairly close to where it is formed, while the protactinium might get carried far away from its point of origin. Thus for "normal" conditions, McManus could establish a ratio of the amount of protactinium to thorium in bottom sediments. When deepwater circulation stops, however, the protactinium will not be carried far away, and thus there would be a higher ratio of protactinium to thorium for sediments formed when the circulation was shut down. McManus likes this proxy because it is less vulnerable than some other proxies to being distorted by an increase or decrease in biological activity.

His first major study looked at an extreme cold event 17,500 years ago, when a massive discharge of icebergs shut down ocean circulation and plunged the North Atlantic into a 2,000-year deep freeze. In that study, published in *Nature* in April 2004, McManus and colleagues (including Keigwin) found that at that time deepwater circulation virtually stopped. They also found a shutdown for the Younger Dryas and correlations between the thorium-protactinium ratios for a number of other abrupt climate changes during glacial times.

Now he is looking at the Holocene, particularly at the times 2700 B.P., 5200 B.P., and 8200 B.P., when the carbon isotope record suggests weaker circulation. When they had looked at long-term data, they found some correlation. Inspired to look closer, McManus says, "Now we're tearing into those same cores, every two centimeters, or roughly every two hundred years, to see whether that relationship bears close scrutiny." He is also looking at cores from a number of different sites in the Atlantic with an eye to creating a continent-sized map of changes in the ocean circulation over time.

The most recent ripple through the Holocene was the Little Ice Age. Depending on which scientist you talk to, the event either ended in the early part of the nineteenth century or still continues today, but the real focus of debate among climate scientists is what caused the cooling. If, for instance, the Little Ice Age involved a shutdown of ocean circulation, it would settle the question of whether such shutdowns have occurred during the Holocene, and it would have very scary implications for those of us living today. As Broecker dryly put it during a conversation, "It would mean that the oceans aren't stable." For one of McManus's colleagues at Woods Hole, Lloyd Keigwin, this question has become something of a quest. And in the summer of 2004, after more than a decade of different probes, he came up with an astonishing answer.

Keigwin works out of Woods Hole's McLean Laboratory. For most Americans, Woods Hole is probably best known as the pretty Cape Cod town that serves as a terminal for ferries to Martha's Vineyard and Nantucket. Camouflaged by picturesque summer homes, however, is arguably the best collection of oceanographic expertise and marine labs on the east coast. Apart from WHOI, Woods Hole is home to the National Oceanic and Atmospheric Administration's Marine Biological Laboratory as well as the private Woods Hole Research Center. WHOI's buildings have two central clusters: one in the village (where the institute also has docking facilities for its research vessels) and another in the nearby Quissett Campus, which contains the McLean Laboratory where Keigwin does his chemical analysis and stores his sediment.

A former Navy officer, Keigwin suffers the misfortune of being prone to seasickness. His subsequent career choice reveals a masochistic streak since collecting and analyzing deep-ocean sediments has condemned him to nausea-inducing research cruises for the rest of his working life. In the course of his career, he has moved from studying ancient forams millions of years old to relative youngsters from the Holocene, prompting an official at the National Science Foundation to joke that the next logical step would be for Keigwin to study the future, at which point NSF would no longer need to fund him. Keigwin plays a crucial role in determining whether there have been Holocene shutdowns of THC because of his reputation, and because during his studies of forams in the Holocene he has produced evidence on both sides of the issue.

Keigwin, like Haug, is looking for a detailed record, which means that he gravitates to places where there is a very rapid accumulation of sediments that would permit him to see distinctions on the short timescales of abrupt climate change. At Woods Hole, he showed me a long sediment core taken from the Bermuda Rise off Cape Hatteras (an attractive area for a paleoclimatologist because it's where the Deep Western Boundary Current runs right alongside the Gulf Stream). The core he showed me looked like a plug of light- and dark-colored mud. The dark areas of the core represent glacial-era deposits when foraminifera were scarce; the whiter areas from warmer times reveal an abundance of limestone residue from the foramanifera skeletons. Keigwin probes these deposits every which way, looking at nutrient levels, ratios of various minerals, and radiocarbon dates, all of which reveal something about what was going on in the oceans at different points in the past. Even with this high-resolution data, however, Keigwin found it maddening to try to determine whether a shutdown of the ocean conveyor was involved in the Little Ice Age or other Holocene events.

The problem is that different tests of the forams extracted from these and other cores yield contradictory results. In one study, for instance, he looked at nutrient levels. The logic was that nutrients levels in the sediments should give a pretty good picture of the conveyor because the water that cools and sinks in the Norwegian Sea does not have much nutrient content. When the conveyor shuts down, southern ocean waters replace the waters coming in from the north, and these waters have a high nutrient content. So high levels of nutrients at some level of the sediments should reveal a shutdown of the THC, and that's exactly what he found when he analyzed his cores.

"I produced that result ten years ago, and it gave me chills," says Keigwin, "because I knew what it meant. I chaired a meeting on abrupt climate change in '95 before the subject became popular, and I knew that the topic was so important that I did not want to go on record and then later have to retract statements about the shutdown of circulation during the Holocene."

So Keigwin teamed up with Edward Boyle, a geochemist at MIT, to analyze the cadmium-to-carbon ratios of bottom-dwelling forams called *Nutallides umbonifera.* Again, Keigwin was looking for nutrients, only this time in the shells contained in the sediments. Cadmium is one of the nutrients these creatures use to build their shells, and in the Rube Goldberg world of proxies, it has surfaced (so to speak) as a proxy for ocean circulation because, as noted, the amount of nutrients vary in the ocean depending on the type and origin of the seawater. Again, as noted, the water that cools and sinks in the Norwegian Sea does not carry much in the way of nutrients. One hundred years later, that water makes its way to the Bermuda Rise, so Keigwin and Boyle reasoned that if they saw a sharp rise in the cadmium-to-carbon ratio in the Sargasso Sea bottom-dwelling forams, the change would indicate that about one hundred years earlier, the deep-water circulation had shut down, shifting the source of water to more southern, nutrient-rich waters. The puzzling result was that Boyle could detect no change in nutrients during the LIA. At this point then, Keigwin had evidence on both sides of the question of whether deep-water circulation had shut down during the Holocene.

Neither Keigwin nor Boyle sees this result as proof that the ocean circulation did not shut down. Rather, the confusing result further underscores a vulnerability of all proxies, which Keigwin terms the "high noise-to-signal ratio." Proxies are generally tiny and sparse signals, and, as noted, it is difficult to know whether the result reflects some unanticipated local effect or global change. So, says Keigwin, with one proxy indicating shut down and another suggesting no change, "I needed a tie-breaker."

He thought that he found it in radiocarbon dating. Deep water ages at rates that can be measured. By comparing the age of the water with stable carbon in the skeletons of animals that live on the bottom, he could determine whether the ocean current above the bottom was stalled or moving. If the radiocarbon changes the same way that stable carbon changes at a given layer of sediment, it would suggest that the deepwater circulation had shut down.

When he analyzed the results from his July cruise, however, Keigwin was gobsmacked. "The result was really perverse," he told me. "I thought that there were two possibilities, but it turned out there was a third—and it's good, shockingly good" (at least for an oceanographer). His results suggested that the vigor of deep-ocean circulation actually increased during the Little Ice Age.

The preliminary result resuscitated an idea that Keigwin had first considered ten years earlier: that the Little Ice Age involved changes in the setup of atmospheric systems over the Atlantic. "This tells us that the ocean could be doing one thing for little cold events, and another for big events," says Keigwin, "and it sets us up to be completely misled when climate changes in the future. Let's say a cooling event started tomorrow, and yet data showed ocean circulation getting stronger. Just as we start to get complacent, bam! It shuts down." Wonderful!

It's also possible, of course, that there are fourth or fifth possibilities that Keigwin and others haven't considered. As he and other scientists point out, we've measured only tiny pieces of the ocean-bottom. A sharp change in ocean-bottom sediments that looks like the record of a dramatic shift in climate might merely indicate a shift in the path of the ocean currents above the bottom. Keigwin tries to guard against this by testing his results against other proxies, but no proxy provides bulletproof evidence of past activities.

Since Broecker had gone into print in 2001 suggesting that the Little Ice Age involved a shutdown of THC, I expected that he would be disappointed by Keigwin's results. But Broecker seems to delight in challenges to theories, even his own. According to Keigwin, Broecker could barely contain his excitement, and offered to help him find immediate funding for further studies. "Republicans would call him a flip-flopper," says Keigwin, "but he's really just a good scientist." Moreover, Broecker is in a position to deliver—to try to speed up the pace of science concerning rapid climate change, Gary Comer, the founder of Lands' End, has made Broecker a gate-keeper to channel the millions he is donating for research. When I ran into Broecker a few weeks later and asked him about Keigwin's results, he said, "That's what science is about—taking measurements and adjusting your thinking."

As McManus's and Keigwin's work implies, the great race to determine the role of the oceans in Holocene climate change continues to gather momentum. There are many other scientists also working on this issue, which has become a hot topic in scientific circles. McManus tells his students that if they want to answer big questions, paleoclimate, not physics, is the place to be.

With the strengths and weaknesses of the science of paleoclimate and its various proxies in mind, let's now revisit some of the examples offered in Part One. Needless to say, there are alternative explanations for each of the calamities and events discussed. How well does the case hold up?

PART THREE

Cross-Examination and Redirect

12

The Mystery of Tell Leilan

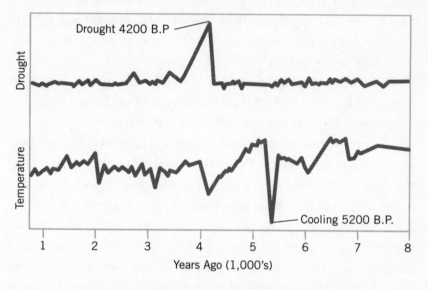

SIGNALS OF DROUGHT FROM MT. KILIMANJARO

MOST OF Harvey Weiss's career as an archaeologist has been given over to the investigation of Tell Leilan on the Habur Plain in Syria. As a student in 1964, he came upon mention of an Assyrian city built on the site while reading an ancient itinerary. By 1979, he and his colleagues had uncovered Assyrian structures, and by the late 1980s he had dug down far enough to find the Akkadian capital beneath the Syrian layer, as well as evidence of its sudden end at around 2200 B.C. At the point at which he began to speculate on whether climate played a role in Akkad's fall, many of the clever proxies used to determine past climate that would subsequently bolster his case had not yet been developed. Similarly, the GISP2 project was still in its planning stages, and the idea of rapid climate change was a wild theory. Thus, when Weiss first put forth his argument, he was somewhat ahead of the really high-resolution scientific data on what was happening in climate 4,200 years ago in Tell Leilan, and why. This would have left him open to criticism in the most genteel debates, but Weiss has a combative personality, and he had just tossed a hand grenade into archaeology, so the debate that he launched was not particularly genteel.

Weiss's theory about the woes of Akkad relaunched climate as a force in history (H. H. Lamb had earlier made broad historical arguments about the role of climate, but Weiss was putting out an example for which peers in climate and archaeology could examine the precise data for his assertions). It's important to keep these external elements in mind. Scientific disciplines have an immune system that initially rejects novel ideas, and archaeology, as John Steinberg noted earlier, is particularly sensitive to explanations that emphasize the role of ecology and minimize the role of human ingenuity. That being said, Weiss would likely have sparked less controversy had he published his first major article in *Science* in 2003 instead of 1993.

The story of Weiss and Akkad offers a fascinating case study in which an explanation of the ties between Akkad's fate and climate became more convincing even as its scientific underpinnings changed and evidence went in and out of vogue. The story also illustrates the science-in-real-time flavor of climate investigations of the past fifteen years.

There's no question that Weiss's tale of the fall of Akkad is a powerful story and, to use the serial-killer analogy, probably sufficient for a conviction in most courtrooms. As with any courtroom drama, however, there is another side. It's open to question whether any evidence would convince the jury of archaeologists that climate is the culprit.

There are archaeologists and other students of the ancient world who challenge almost every detail of the narrative of Tell Leilan offered in Chapter 4: from the reason for the collapse of Tell Leilan (it wasn't climate that did the Akkadians in, the alternative story goes, but invasion, or the political disintegration of an overextended empire), to whether there even was an abrupt shift to long-term drought in the region, to whether other contemporary collapses cited by Weiss in Harappa and Mohenjo-Daro really took place, to whether weather played a role in the end of the Old Kingdom in Egypt, and on and on. Weiss, the most passionate proponent of the role of climate in the synchronous collapses, had the boldness to contradict the conventional wisdom of a hidebound archaeological community, but he seems to also relish battles.

In fact, there are at least three related stories about the events surrounding 2200 B.C. One is what actually happened. Another is the changing body of evidence that supports accounts of what happened during that time. In the last few years, Weiss has shifted from embracing as compelling one piece of evidence in the paleoclimate record to declaring it irrelevant. And third is the story of how the rapidly unfolding science has reverberated through the archaeological and paleoclimate communities.

So let's go back through this story again, this time replaying crucial aspects of Weiss's argument against the conventional wisdom in the archaeological community and the state of knowledge of past climate.

When Weiss was getting his Ph.D. in archaeology at the University of Pennsylvania in the mid-1970s, questions of climate rarely intruded when it came to reconstructing the fortunes of past civilizations. One of the defining characteristics of a civilization is that its institutions and infrastructure—

banking, granaries, the rise of trades and skilled labor—are supposed to insulate a people from the vagaries of the weather. Perhaps most important, at the time when Weiss began his career, most paleoclimatologists ignored the Holocene as a monotonously stable period.

Thus, when Weiss joined the Yale faculty and went off to Syria as research director of the Tell Leilan dig, the notion that he would unearth evidence of one episode of a global apocalypse was nowhere in his mind. But the soils, seeds, and other detritus he was excavating told such a compelling story that he was forced to think beyond the conventional explanation that the empire was overextended ("too many mergers and acquisitions" was the way he described the consensus description of Akkad's fall in a 1996 article in *The Sciences*).

In a part of the dig he labeled Trench B in the southern part of the lower town, Weiss and the team uncovered what he analyzed as three distinct phases. The first phase contained evidence of human trampling, some volcanic residues, some dissolved mud brick, and some indicators of earthworm activity—a signal that soil moisture was still relatively high. Things changed dramatically in the layers above phase I. The next layer consisted of about 20 centimeters of various types of sand and silt characteristic of windblown deposits. There was little evidence of earthworm disturbances. Upon analysis, Weiss thought that he was looking at indications of drought interrupted by episodes of erosion of a bare surface soil by violent storms. As he and six colleagues wrote in their 1993 article in *Science,* "Both soil and sediment features may be evidence of the establishment of marked aridity induced by the intensification of wind circulation, and an apparent increase of dust veil frequency" compared to the present. By phase III, the rains, people, and earthworms returned.

For Weiss, it seemed unlikely that political disintegration could cause people to abandon one of the more productive areas of the Fertile Crescent for three hundred years. Volcanic ash from an eruption might have temporarily disrupted climate, but, again, any such disruption short of a catastrophic event would be limited to a few years. So he had to look for other causes.

Even before he published the *Science* article, the sharp line in the sand revealed by the dig at Trench B caused Weiss to look at a variety of clues in a new light. There was the text of "The Curse of Akkad," which most archaeologists dismissed as poetry rather than as a historical account. There were also references to drought and suffering in some cuneiform tablets. The story of the soils caused Weiss to reconsider some snippets from the archaeological literature that had not registered earlier. He notes that Robert McAdams, the former director of the Smithsonian Institution, had glossed over a "short, major, wind-erosion cycle" in the area at the time of the collapse, and also remembered that a British archaeologist, James Mellaart (who first achieved fame for discovering the world's first city in Catalhoyuk, Turkey), had earned catcalls in the 1960s for citing drought as the culprit in the collapse of many of the same civilizations and cities that later caught Weiss's attention.

Then there was the unfinished wall described in Chapter 4. Weiss uncovered it in 1993. He argues that the wall offers a crucial clue to the fate of the Akkadian empire. The clue, says Weiss, lies in what wasn't there. When the dig team first exposed this part of the site, the archaeologist was puzzled because there were no bricks to be found. "I'd spent years digging in this area," he recalls today, "but never had I come upon basalt with no brick. My first thought was, 'Who did this pilferage?'"

In 1999, Weiss returned to the partially uncovered site, still puzzled. Then, driving across the baking plain back to his quarters after studying the excavation, he suddenly realized what was staring him in the face. He slammed on the brakes as he realized what the basalt was telling him: *"They never finished building the wall!"*

From that point, he systematically began to test this hypothesis. Adjacent to the dressed basalt blocks, the team found a completed wall with a basalt base, a mud layer, potsherds, and then topped by four courses of bricks. Around the area, they found basalt blocks in various stages of dressing. As they painstakingly removed the entombing dirt, they uncovered the tableau of a drama that had been frozen in time for 4,200 years. The scattered blocks and unfinished wall suggested that workers had stopped work in their tracks and left. "Someone gave the order," he notes, "and they moved out, probably in a matter of days."

When Weiss published his hypothesis in 1993, the corroborative evidence for reconstructing ancient climates was relatively sparse, and, as we have seen, the notion that climate might change abruptly was only just beginning to dawn on scientists. But in 1993, he did have access to ice core data from an earlier probe of Greenland. He also had access to some lake-bed sediments from the area and to the data drawn from a limited sample of tree rings collected from ancient timbers from Anatolia.

With only an incomplete mosaic of hard data to support his contention of a major climate upheaval 4,200 years ago, Weiss immediately found himself on the defensive as the archaeological community reacted to this impudent repudiation of the conventional wisdom. Prominent among his attackers was Karl W. Butzer, a specialist on the interaction of environment and archaeology at the University of Texas. Weiss invited Butzer to air his dissent in a volume Weiss coedited entitled *Third Millennium BC Climate Change and Old World Collapse,* and Butzer did not hold back. He cited analysis of lake-bed sediments, Dead Sea samples, and other proxies to suggest that even if climate changed, it was not a regional change and it was not abrupt: "There was, then, no abrupt climatic shift to greater aridity affecting the larger region—the Near East and the Aegean world—between c. 2400 and 1900 B.C.E.," he flatly asserted. He also claimed that even when Tell Leilan collapsed, other settlements on the Habur Plain continued to do just fine. Even if it proved true that Tell Leilan suffered poor harvests, Butzer suggested, its residents would have had access to grain produced elsewhere through trade relationships. He saw the collapse of Tell Leilan as a political event following a breakdown of the market system that governed the region's economy. If climate was a factor, he asserts, it was but one factor among many.

It's understandable that Butzer would take issue with the elevation of climate above other factors in a society's collapse. He had built a distinguished career articulating a perspective that views civilization as an adaptive system that is constantly and unpredictably integrating a panoply of selective pressures, ranging from external events such as drought and invasion, to internal economic and sociopolitical challenges. Collapse can occur because of a chance concatenation of events, or, as he wrote in 1982, ". . . the unexpected coincidence of poor leadership, social pathology, external political stress, and environmental perturbation can trigger a catastrophic train of mutually reinforcing events that the system is unable to absorb."

For Weiss, on the other hand, the question of adaptation is a canard: "They did adapt; they left," he explains simply. "Habitat tracking [which means, in essence, moving to where the grass is greener] is a fundamental cultural adaptation to conditions that can't sustain life. Adaptation does not mean staying in one place regardless of what happens."

While the debate will continue over whether climate was a determinative factor or but one of many shocks that contributed to the fall of Akkad, the evidence available to test Butzer's contention that climate did not change abruptly and over large areas in 4200 B.C. has increased dramatically in the years since Weiss first published his hypothesis in *Science*. The 1990s saw an extraordinarily rapid advance in the understanding of past climates, and this advance in understanding the past precipitated a dramatic shift in the paradigm of how climate changes. As Peter deMenocal put it, "When I began my Ph.D. in 1986, the conventional wisdom was that it took one thousand years to end an ice age. By the time I finished in 1991, that figure had been reduced by an order of magnitude to one hundred years. Just two years later, Richard Alley showed that climate could change from warm to glacial conditions in two to five years."

Weiss had no idea of the seismic rumblings in paleoclimatology just beginning to roll out of Greenland when he published his paper, nor did Mayewski know that Weiss and other scientists were looking for stronger data of a global climate change 4,200 years ago. Then, two years later, a couple of chance encounters opened whole new worlds of data for Weiss, and at first he embraced GISP2 as a drowning man would a spar.

In the summer of 1995, Paul Mayewski's assistant came across Weiss's 1993 article in *Science* and sent a note to the archaeologist mentioning that the 4200 B.P. climate-change event might be in the Greenland record. Weiss got the note just before he left for his vacation in Maine. He looked up Mayewski's number (Mayewski was then at the University of New Hampshire), and, when he got him on the phone, he jumped right in. Introducing himself, he said, "Do you have this? It's amazing if you have this."

Mayewski looked at his data and indeed he did have a signal of the 2200 B.C. event—at least to his satisfaction. Mayewski claims that the event shows up very clearly in a number of individual proxies taken from the ice cores, as well as in a statistical function he developed that merges all the various proxy records into an "empirical orthogonal function." Mayewski also asserts that there are clear connections that link the events revealed by these proxies to climate changes in western Asia.

One of the proxy records, for instance, details the varying amounts of sea salt that has been laid down annually on the ice. The sea salt gets blown onto the ice sheet by winds coming off the North Atlantic, and Mayewski argues that the variance reveals the relative storminess of the Atlantic and the relative amounts of sea ice. More sea salt equals more storminess and sea ice; less sea salt correlates to reduced winds and more summery conditions. Another proxy is dust which, Mayewski contends, also gets blown onto the ice sheet when winds are more intense. The smallest level for sea salt in the Holocene shows up at 2200 B.C. in the GISP2 record, when dust was lower as well.

Mayewski contends that this implies that at that time, atmospheric circulation was less vigorous during the winter and spring periods when Mesopotamia ordinarily would receive the bulk of its annual rainfall—moisture that originates in the North Atlantic and the Mediterranean, and the reduced seasonal winds would lead to drought. Notably, the lowest values for dust and sea salt took place in the span between 2200 B.C. and 1900 B.C., the identical dry period that Weiss could see in the soil profile at Tell Leilan.

Mayewski argues that Akkad was a victim of one of the regular rapid climate cycles that ripple through the glacial era and continue in the Holocene. In this case, Mayewski thinks the likely culprit is a regular 1,450 cycle. The cycle shows up in the GISP2 record, but Mayewski is not sure whether it results from an oscillation in ocean circulation, changes in the amount of energy delivered by the sun, or some other cause.

When I first spoke to Weiss and Mayewski in 1997, they were embarking on a collaboration that extended far beyond

Mesopotamia. They had joined forces to explore links between events revealed by the ice cores and the collapse of several other civilizations. The two scientists got as far as submitting a paper to *Science* that detailed these connections, but some of the respondents in the peer review process raised the question of whether the authors could establish that teleconnections really linked Greenland climate changes to events in western Asia and other temperate, subtropical, and tropical regions.

Seven years later, Weiss seems to have joined the camp of these critics. "What you see when you look at the spikes in the GISP2 record for 2200 B.C.," he said in 2004, "is a statistically massaged record. The empirical orthogonal function [Mayewski's equation that merged various records] is a multivariant analysis that throws all these proxies together and comes up with a number. If you look at the individual proxies, you don't see it." Such statements are a complete turnaround from Weiss's position of just a few years earlier. Moreover, he is now more skeptical of the evidence in GISP2 and its teleconnections to western Asia than some of the most respected names in the climate community.

Jeffrey Severinghaus of the Scripps Institution of Oceanography, for instance, believes that the 2200 B.C. event shows up quite clearly in GISP2. "If Weiss states otherwise, he's just wrong," says Severinghaus. While Alley notes that the record of 2200 B.C. is not as striking as some of the other signals from the Holocene ice cores, he accepts Mayewski's assertions, saying, "Paul does not make things up."

Whether the Arctic was driving climate events at that point or recording changes is another question. Alley argues that Greenland's utility as a record of far-flung change is incontrovertible if only because many of the traces entombed in its ice can be tracked back to far points on the globe. As a driver of climate, however, Alley acknowledges that the North Atlantic might have moved from leading to supporting role in the Holocene as its reservoirs of ice diminished. Glacial ice and sea ice play a crucial role in the shutting down of ocean circulation and the amplification of those effects to the point that they can have a large impact on climate. "In the Holocene, there is less connection between the monsoon and events in Greenland than there was during the Ice Age," Alley says. "As ice gets smaller, it's possible that the North Atlantic has a harder time reaching out to the rest of the world." Others continue to see a connection.

Whatever the merits of Weiss's dismissal of the ice core corroboration of his finding in Tell Leilan, one reason he can afford to be blithe is that since he turned to Mayewski for hard data, dozens of other records have surfaced or been developed that confirm that 2200 B.C. was a period of drought in the Middle East and dramatic climate change around the world. One particularly relevant record grew out of an encounter between Peter deMenocal and Weiss in an unlikely setting: a NATO-sponsored conference organized to explore the linkages between climate and civilization collapse in the ancient Middle East.

At the conference, deMenocal listened to Weiss and started thinking that if the events Weiss described had happened, there should be some record of them in seabed sediments in the Gulf of Oman. Even though it is 2,200 kilometers from Tell Leilan, deMenocal thought the wind patterns prevalent in dry months would carry dust from Mesopotamia south and east over the Persian Gulf to the Gulf of Oman. Later he obtained satellite photos that documented movements of one dust storm that bore out this hypothesis. DeMenocal had earlier ignored the Holocene because he shared the prejudice that it was stable and boring, but at the conference he approached Weiss and said, "If you're looking for hard data, I can get it for you."

Teaming up with Heidi Cullen, then a Lamont-Doherty–based expert on ancient sediments (and later an on-camera climate expert for the Weather Channel), deMenocal and his team took a detailed look at a 6-foot core previously extracted from the Gulf of Oman by a German paleoceanographer. They looked for levels of dolomite, quartz, and calcite, characteristic residues of Mesopotamian dust. They narrowed the origins of the dust by identifying chemical tags for different regions, and they narrowed the hundred-year-plus margin of error of carbon-14 dating techniques by ground-truthing their results against tephra laid down by specific volcanic eruptions in the region. The results, presented at the annual meeting of the American Geophysical Union in the fall of 1997, suggested that around 2200 B.C. the amount of dust increased to levels not seen since the last ice age, an indication of extraordinary cooling and drying. Moreover, the spike in dust persisted for a few hundred years. The analysis brought around a few formerly dubious members of the archaeological community, but even this record came under scrutiny.

Tony Wilkinson of the University of Chicago's Oriental Institute wrote a letter to *Science* following a news report of deMenocal's findings in which he suggested that the dust might have come from Yemen or other parts of the Arabian Peninsula, not Mesopotamia. He also supported Butzer's contention that the Akkadians had adaptations to absorb the shock of climate events. By this time, however, every year brought more supporting data of both a generalized climate change in 2200 B.C. and specific evidence of cooling and drying in the region.

Lonnie Thompson of Ohio State University first found evidence of the 2200 B.C. event in glacial ice in Chile that suggested that the Amazon suffered the worst drought in the previous 17,000 years at the same time. In the winter of 2000, he and a team extracted six cores from the fast-shrinking glaciers atop Mount Kilimanjaro on the border of

Kenya and Tanzania in East Africa. That record showed a huge dust spike around 2200 B.C. that lasted for three hundred years. Study of lake-bed varves from Elk Lake in northeastern Minnesota recorded the big climate shift of that time. At the same time, new records were becoming available much closer to Mesopotamia.

Weiss now details forty-one climate proxy records from areas in Africa, Europe, and the Middle East that support the contention of a dramatic drying and cooling at the time of the collapse of the Akkadian Empire. Apart from marine, lake, glacial, and dendrochronological records, there is also the climate record reconstructed from analysis of stalagmites and other forms of speleothems in regional caves. Weiss argues that this record offers the highest resolution of any record yet available (though the ice core advocates disagree strenuously).

The slow drips of water that create these fantastic cave structures leave behind annual bands whose thickness and chemical makeup can be examined. Again, the researchers are looking for ratios of O_{18} to O_{16} and other isotopic records to reconstruct past climates. The sheltered nature of cave deposits makes them particularly attractive to paleoclimatologists. Miryam Bar-Matthews of the Geological Survey of Israel has studied speleothem deposits in the Soreq Cave and other places in Israel and reconstructed a 200,000-year time line of climate for the region. Again, the 2200 B.C. event shows up very clearly.

Given the accumulating evidence, as well as the accelerating rate of that accumulation, it seems likely that Weiss's argument holds up against the criticism that there is no compelling evidence of protracted drought in the area of Tell Leilan commencing about 2200 B.C. By 2100 B.C., temperatures in subtropical and subpolar regions dropped between 1 and 2 degrees centigrade on an annual basis, and precipitation in Mesopotamia may have fallen by more than 50 percent.

Moreover, the signals of abrupt climate change show up around the world, suggesting that the events that made life difficult on the Habur Plain also impacted a number of other cultures and civilizations at that same time, a claim that will be taken up in subsequent chapters. In fact, those people who were perhaps too quick to dismiss the evidence of a global signal of climate change must include Harvey Weiss. It has become clear that the events that left an imprint on Kilimanjaro, cave formations from Israel to China, and deep-seabed sediments from Venezuela to the Gulf of Oman also left their imprint on the ice laid down in Greenland.

The other major criticism—that the fall of Akkad may not have directly resulted from the protracted drought—is more difficult to answer. While Weiss contends that the area became all but uninhabitable, there are others who counter that there is evidence of uninterrupted habitation in adjacent areas. If, however, the Akkadian Empire was leveraged on past patterns of rainfall, it is very easy to envision that a drought could very quickly make it insolvent by denying its masters the currency—grain—that they used to pay their laborers and warriors. Moreover, if Akkad came apart because of political or social upheaval, it seems likely that some group would have taken advantage of the power vacuum to seize Tell Leilan in the aftermath. That did not happen for about three hundred years, suggesting that the city was uninhabitable or cursed, as the ancient poem would have it. Weiss makes a strong argument that drought was the curse of Akkad.

The story of the fall of Akkad is not a perfect story of the connection between climate and the fall of a civilization. Richard Alley, for one, says that if you began with the ice core record of events and then looked for historical upheaval, it's open to question whether you would start with 4200 B.P. "Once Peter deMenocal says, 'Take a look,' however," says Alley, "you see it." Still, it was the first story that brought the linkages between climate and history to the attention of a generation of climate scientists and to the American public. H. H. Lamb had earlier made similar arguments for other civilizations, but, farsighted though he was, he lacked the hard data that became available as Weiss pursued his case.

There are a number of scientists who still regard both the proxy evidence and the arguments that link Akkad's fall to climate as speculative, but some of these same scientists accept the linkage between climate change and the fall of the Mayans, for instance, or the links between climate and the end of the Greenland colonies. Even though the supporting evidence changed from ice cores in Greenland to speleothems in Israel, and even though Weiss might have amplified the contentiousness of the debate about his contentions, he deserves a great deal of credit for breaking down the barriers that previously confined climate scientists and ancient historians to parallel universes that rarely intersected.

What happened to the world's climate 4,200 years ago remains something of a mystery. Whether climate also cooled 4,200 years ago is open to question, as is the issue of whether the dramatic events were the result of a regular climate cycle or instead reflect some rogue wave in climate that emerged from the harmonic convergence of a number of different factors that impact climate. Richard Alley refers to such events as the result of "stochastic resonance." As he explains it, "The sun might be in the background quietly saying 'Get warmer or get colder.'" The world is a noisy place, he continues, and the sun's message might be amplified or contradicted by other messages so that a 1,450-year cycle might occasionally show up stronger or become so weak that it appears to have a 3,000- or 4,000-year spacing.

Lonnie Thompson sees the traces of extreme drought in huge spikes of dust in tropical glaciers around the world. The change was also abrupt, but was it an abrupt climate-change event? Richard Alley analogizes abrupt climate-change events to tornadoes in weather and defines them as threshold-crossing events in which changes occur faster than their cause.

Whatever the cause of the drought 4,200 years ago, Weiss seems to be gaining converts. One powerful ally is Ofer Bar-Yosef of Harvard, whose specialty is the interplay of environment and material culture. He says that from his earliest days studying in Israel he believed that environment and climate played an important role in the life of complex societies. He mostly kept these thoughts to himself, however, because anything relating to environmental determinism was regarded as suspect in American academic circles. Then he came across Harvey's story of Akkad based not just on proxy records, but on hard physical evidence. "It's a very good story," says Bar-Yosef, noting that after thirty years of biting his tongue, "Harvey made me feel comfortable that my early instincts were justified."

13

Scorched Earth

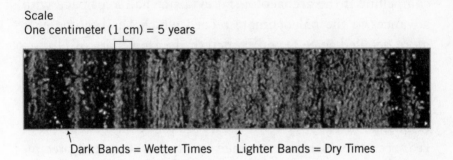

Dark Bands = Wetter Times   Lighter Bands = Dry Times

LAKE BOTTOM SEDIMENT CORE FROM YUCATAN PENINSULA

THE COLLAPSE OF the Mayans has spawned over a hundred different theories of the civilization's fall. Climate is a relatively recent entrant, following other ecological factors such as deforestation and overpopulation, but here too, as in the case of Akkad, the pace of scientific discovery has played a major role in the debate. In just the space of a decade, between 1993 and 2003, the resolution of proxy records increased by an order of magnitude.

This is not to say, however, that resistance to such linkages has collapsed. As Lisa Lucero puts it, "Many Mayan experts endorse the notion that the Mayans were special, a unique civilization, and because of this there is the feeling that the reason for the collapse must be special as well." Many anthropologists resist using the word "collapse" when discussing the fate of the Mayan civilization. In the north of the Yucatán, the culture persisted into colonial times, even if the signature buildings, high science, and Mayan monuments disappeared into the jungle. Lisa Lucero argues that there is no contradiction in the rise of some secondary cities during the time Tikal collapsed. A number of those surviving centers had local water supplies and fertile land.

Lucero acknowledges that her scenario would be more compelling if the archaeological evidence had kept pace with advances in the paleoclimate record which has become ever more detailed since David Hodell of the University of Florida found evidence of protracted drought at the time of the Mayan collapse. Hodell advanced this idea after examining lake-bed sediments from Lake Chichancanab in the Yucatán Peninsula. In 1993, he retrieved a core that contained a 9,000-year record, which he examined for temperature and precipitation data. His proxy for precipitation was gypsum that flowed into the lake via streams and creeks. During wet periods, there was enough water to contain the gypsum in solution and most deposits fell to the bottom near the shore. During dry periods, however, water volume shrank and the lake would quickly become saturated with gypsum, with the consequence that gypsum would fall to the bottom throughout the lake, even in its deepest parts near the center. Thus, Hodell reasoned, the relative size of gypsum layers at the center provided a proxy for drought—the more gypsum, the drier the weather that year.

That first core covering a 9,000-year span did not have the resolution for Hodell to see anything more specific than relative dry and wet periods. One of those dry periods, however, started around A.D. 750, and continued on and off for more than two hundred years, right around the time of the final Mayan collapse. It was electrifying news. Even the most skeptical archaeologist would acknowledge that a more than two-hundred-year dry period could do in even the most robust civilization, particularly since the dry period came on the heels of a long stretch of good weather during which the Mayans greatly expanded their numbers and overtaxed the land. On the other hand, the finding raised as many questions as it answered, particularly concerning the ubiquity and intensity of the event, and whether other factors besides drought might have caused variations in sedimentation. For instance, some critics speculated that erosion as a result of deforestation might have torqued the rate at which sediments flowed into the lake.

Other records filled out the picture of drought. Hubert Robichaux, an anthropologist at the University of the Incarnate Word in San Antonio, found evidence of thirty-two years of reduced flow in the Río Candelaria in Mayan Mexican territory during the time of terminal collapse. Richardson Gill's compared the paleoclimate record with the last recorded dates on monuments in various Mayan centers. He concluded that the collapse occurred during three distinct phases starting in A.D. 760 and ending in A.D. 910. These dates coincide with intense periods of drought.

Then in 2003, Gerald Haug and colleagues published their study of seabed sediments from the Cariaco Basin off Venezuela. While not in the immediate environs of the Mayans, the basin was subject to the same movements of the Intertropical Convergence Zone. More important, the particular core they looked at offered a record of change on an annual basis, more than ten times the resolution of Hodell's original lake-bed record. Given that the land surrounding the basin was sparsely populated, there was less chance that sediment changes were altered by land practices such as tree felling or other factors that spur erosion.

Haug and colleagues looked at titanium as a proxy for precipitation because it flows into the basin at different rates depending on the amount of fresh water coming in from rivers. The core they looked at had been extracted in 1996 from a depth of 893 meters and contained sediments accumulated at the rate of 30 centimeters per 1,000 years. Using an X-ray fluorescence technique newly tuned to examine sediments in layers as thin as 50 microns, the scientists took a closer look.

They discovered that the years between A.D. 550 and 750 were particularly wet. After a period of decline called the "Hiatus" between about 536 and 593 (perhaps related to the mysterious cold event of 536), the Mayan population exploded, setting up the civilization for later disaster. Then beginning in 760, the scientists discovered peaks of intense aridity, including a severe drought in A.D. 810, a three-year drought that began in A.D. 860, and a six-year drought beginning in A.D. 910. The droughts corresponded precisely with the terminal dates recorded by Gill. Moreover, the sediment record documented that the series of droughts occurred against a general period of low precipitation that made the period starting in 760 the driest in Mesoamerica of the millennium. In the early centuries of the next millennium, the region suffered even drier conditions, but by then there was no Mayan civilization left to kill.

Work remains to be done. While laminations in the cores permit Haug and colleagues to state with certainty that the interval between two droughts might be fifty-one years, there is ambiguity about the absolute dates. Tropical tree growth does not permit the construction of a tree ring chronology that might correlate with the sediment record.

While Haug was looking at seabed sediments, Hodell went back to Lake Chichancanab and in 2000 extracted a new core with an eye toward getting a more detailed look at climate variation. Not far from the place where he had extracted the 9,000-year segment he had examined in 1993, his team found a 2,600-year record that permitted closer analysis because sedimentation rates were 40 to 80 percent higher than the 1993 core. The analysis, which put bits of charcoal, seed, and wood fragments through an accelerator mass spectrometer to establish C14 dates, had a much higher resolution than his earlier work on the 9,000-year section. Hodell and his colleagues analyzed the core at 1-centimeter intervals, sufficient detail to give them a picture of climate at about 11-to-14-year intervals throughout the core. To get temperature data, they examined the O_{18}–O_{16} ratios in the shells of mollusks recovered from various depths, and once again they used gypsum as a proxy for relative aridity.

The results confirmed their earlier findings as well as Haug's. At key inflection points in Mayan history, collapses occurred at the time of severe drought. Hodell's analysis suggested that the arid period, which began in 750 and continued for about 200 years, was probably the driest stretch in that region for the previous 7,000 years. Mayan civilization, which somehow had managed to regroup after the droughts of the Hiatus, had neither the infrastructure nor the explanations sufficient to deal with the much more severe droughts that came later. Moreover, the pressures of increased human numbers on the surrounding land and forest may have left the Mayans even more vulnerable to drought than they had been 400 years earlier.

Hodell published his findings in *Science* in May 2001 and offered a contributing factor in the dry spell. Paleoclimatologists use C14 as a proxy for changes in solar intensity because the isotope enters the biosphere in the form of cosmic rays. Confirming their findings by comparing the Lake Chichancanab data with earlier analysis of tree rings in Sweden, Hodell found that the arid periods correlated with a 206-to-208-year cycle in solar energy. He noted that at the same time that the Mayans were baking under drought, so was the Sahel in sub-Saharan Africa. Even more than the Yucatán, the people of the Sahel live by the vagaries of the Intertropical Convergence Zone.

Something shifted this zone southward across much of the earth. Since the zone lies in the part of the planet that receives the most intense solar energy (and produces the most evaporation), a diminution in solar radiation might reduce the movement of the ITCZ from its equatorial anchor. In *Science,* Hodell acknowledges that civilization-killing droughts don't occur every two hundred years in Mesoamerica (and that the Mayans had survived many previous bicentennial solar cycles), so clearly other factors were at work as well. Perhaps this is another case of the stochastic resonance proposed by Richard Alley, a situation where the sun is saying "Get cooler" in the background, while other factors contribute to a type of harmonic convergence of drought-enhancing effects.

If Hodell is correct, there may yet be a lesson of comeuppance in the fall of the Mayans. The pride of their science was astronomy, particularly keyed to seasonal movements of the sun. They prospered by using that knowledge to perfect an agriculture that could extract a living from harsh conditions. He notes that the Mayan calendar took note of cycles of 11, 88, and 212 years—periods that correspond to known solar cycles. While the Mayans might have known about the long solar cycle that possibly did them in, either they could not or did not prepare for its effects. A civilization whose leaders asserted that they had some connection to the forces that controlled the heavens may have been hustled offstage with the help of the sun.

14

Is It Little Ice Age, or Ages?

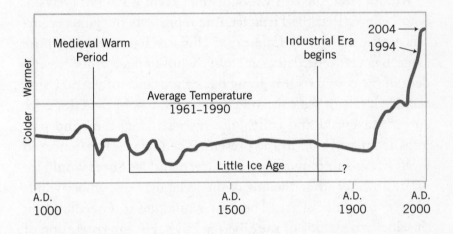

A MAJOR POINT of contention about the Little Ice Age is whether it was a discrete event. "Little Ice Age" covers a five-hundred-year span. According to some climate historians, that may represent several distinct events rather than one continuous episode. Scholars date the LIA from as brief as 1550 to 1700 to as long as the fourteenth century to the early nineteenth. At one extreme, Paul Mayewski argues that the atmospheric setup of the LIA continues today. At the other, Astrid Ogilvie, of the Institute of Arctic and Alpine Research in Boulder, Colorado, would like to discard the term Little Ice Age, with the exception of its use "as a relatively small hemispheric-scale temperature excursion."

Those who argue against a single event cite the fact that the LIA began at different times in different parts of Europe. The first frigid breath hit Iceland early in the thirteenth century, while Italy escaped the brunt until the fifteenth century. Moreover, many different factors could have caused cooling at times during the span of the LIA. It was a time of heightened volcanic eruptions, and the coldest period—the Maunder Minimum—was a time of dramatically reduced sunspot activity, which has also been correlated with cooling.

Arguing that the LIA was a distinct event is Paul Mayewski, who believes that the Little Ice Age represents the latest recurrence of a millennial-scale rapid climate-change event whose cause has yet to be determined but whose fingerprints are evident in the proxy record going back thousands of years. Using relative levels of sea salt and continental dust in the GISP2 ice cores, Mayewski and colleagues reconstructed the wind patterns from A.D. 600 to the present. The logic is that increased levels of sea salt blown onto the Greenland Ice Sheet would indicate a winter strengthening of the Icelandic Low, whose winds blow counterclockwise. The dust would get to Greenland on the clockwise winds of the Siberian High. His interpretation of the record indicates a sharp intensification of the low during the winter and the high during the spring, projecting that the LIA was an era of consistent storminess. Because the intensity of these systems has not yet returned to pre-LIA levels, and because he believes the LIA is the latest instance of rapid climate-change events that typically run longer than 1,000 years, Mayewski says the Little Ice Age has not yet ended. In his book *The Ice Chronicles* (coauthored with Frank White), Mayewski writes that if the LIA is over "it would also be one of the shortest [rapid climate-change events] in the last 110,000 years."*

*Paul Mayewski and Frank White. *The Ice Chronicles: The Quest to Understand Global Climate Change* (University of New Hampshire Press, 2001).

He attributes the warming since the beginning of the nineteenth century to the huge release of greenhouse gases beginning in the Industrial Revolution and accelerating in the twentieth century.

Part of Europe's misfortune during the Little Ice Age derives from its geology. There is an asymmetry in the present-day world that leaves Europe vulnerable to outsized effects from shifts in climate. At the South Pole, there is land on which to build an ice sheet, which makes it easier to store cold in the southern polar region. At the North Pole, there is open water for 10 degrees latitude in all directions, which makes it easier for the north Arctic to shed cold. The south polar region is thermally isolated by an ocean, while in the north heat from the south can regularly penetrate northward and cold can spill southward. The response of sea ice and permafrost can amplify and broadcast the effects of these heat transfers. Europe gets much of its weather from west to east, planetary winds that in winter deliver heat stored in the oceans, from the storm track over the Atlantic, and from the winds and currents emanating from the Polar Regions. All of these are subject to variation. For Europe a certain amount of instability is baked in the cake.

Whatever the cause of the five-hundred-year cooling of the Little Ice Age, one agent directing its storms and freezes was the North Atlantic Oscillation (NAO), an appellation that describes the relationship between the Icelandic Low and the Azores High. Think of the two systems as yet another set of giant gears in the clanking climate machinery, with the Icelandic Low, as always, rotating counterclockwise and the Azores High rotating clockwise. The effect of the two gears is to push the air between them eastward, defining the storm track over the Atlantic Ocean with the exact direction and the intensity of the winds determined by the relative position of

the two systems and the size of the pressure differences between them. If there is very low pressure in the Icelandic Low
and very high pressure in the Azores High, the difference between the two amplifies the intensity of storms.

In concert with the ocean currents, storms deliver heat
away from the equator. When the sun evaporates water from
the surface of the ocean, a good portion of the heat spent to
evaporate that water is retained in the water vapor as it rises
to form clouds. The clouds then travel along the storm track,
and as the vapor condenses to form raindrops, the stored heat
is released into the atmosphere. In this fashion, Europe gets
about half the heat it receives courtesy of the tropics.

How much heat comes north and where it arrives in Europe depends on the storm track, which is determined by the
relative positions of the two pressure systems. This, in turn,
depends to some degree on the boundary between cold and
warm waters in the Atlantic. During positive phases of the
NAO, cold waters between Greenland and Europe remain far
north, allowing the storm track to deliver rainfall and warmth
to the United Kingdom and northern Europe. During negative
phases, when the pressure differences weaken or even reverse,
cold water pushes south and the storm track moves south as
well. During the negative phase of NAO, high pressure can
form over Greenland, setting up a clockwise circulation that
draws frigid air down from the poles to Europe from Siberia.

During the Little Ice Age, both phases of the NAO seem to
have intensified, producing champion storms and numbing cold
spells, which imposed enormous costs on Europe and North
America. One casualty of the extreme weather was the Spanish
Armada, dogged by intense storms on its advance toward the
British Isles, and again during its retreat to the Bay of Biscay.
H. H. Lamb notes that southward-advancing polar water
dropped water temperatures by 5 degrees centigrade between
the Faroes and Iceland the summer the Spanish Armada made

its way north. This set up a tightly packed thermal gradient between about 50 degrees north and 60 degrees north, which in turn fueled what Lamb calls the most intense storms of modern times. Apart from devastating the Armada, other storms breached dikes in Holland, drowning tens of thousands, and pummeled cities in England and on the Continent.

While the question of what caused the Little Ice Age remains open, there is mounting evidence that wintertime temperatures in Europe were much more severe than the global signal of a one-degree cooling might suggest. Peter deMenocal arrived at that suggestion from sediment cores taken from a seafloor slope just off of Nova Scotia as part of his effort to probe climate change in the Holocene. He explains the logic train that leads from seabed sediments off Nova Scotia to temperature differences in Europe as follows:

The deep seawater that sinks in the Labrador Sea forms in vintages. These pulses have distinctive salinity and temperature signatures (signified by O_{18} to O_{16} ratios) that become encoded in the calcium carbonate skeletons of foraminifera. DeMenocal is particularly interested in a surface-dwelling creature called *Globigerinoides,* which extracts materials from seawater to build its shell. DeMenocal and his colleagues dissolved the sand grain–sized shells extracted from a particular point in the sediments in an acid to liberate the CO_2, and then separate the oxygen so that they could measure the oxygen-isotope ratios. Among other things, their results reveal that big changes can be concealed by what look like small variations in annual global temperatures.

When they looked at forams, they found that ocean temperatures varied about .5 degrees centigrade on a multidecade timescale. When they studied the oxygen ratios for forams buried during the Little Ice Age, however, they found temperature differences of three to five times that historical average. This makes for an interesting contrast with the conventional figure for the 1-degree average global temperature change during the Little Ice Age. What this means is that the moderate 1-degree cooling of the LIA camouflaged a much more severe cooling of up to 3 degrees centigrade in the North Atlantic. Given the havoc wreaked by the less than 1-degree temperature change of a strong El Niño, it is easy to see how such a temperature change would be "a very big deal," as deMenocal put it, if it happened today, or when it happened in the past.

DeMenocal's startling finding about the amplitude of temperature changes in Europe (and the surrounding land) during the Little Ice Age reaches the same result as clever detective work done by Richard Alley and another pioneering paleoclimatologist named George Denton, who have been taking summer research trips to Greenland financed by Gary Comer, the billionaire Lands' End founder, who is deeply concerned about the possibility of rapid climate change. Examining the detritus of ancient glacial moraines, they discovered that the debris left by the advance of ice during the Younger Dryas 11,500 years ago was only a few hundred yards away from the present-position glacial terminus, implying that the era's temperatures were not terribly different from today's. Temperature readings from the surface at the center of the ice sheet, however, told a different story, revealing an annual difference of 18 degrees centigrade colder. Why the discrepancy?

Alley and Denton reasoned that the small difference in glacial moraines was telling them that summer temperatures remained similar to current levels since colder winter temperatures would not affect the edge of the glacier (once it is below freezing, further drops in temperature do not appreciably increase snow accumulation), but warm summer temperatures would. That means that winter temperatures must have been much, much colder to get to the annual average of 15 to 18 degrees colder than the present, perhaps as much as 27 degrees centigrade colder. The Little Ice Age reveals the same discrepancy, suggesting that while summer temperatures did not change that much, wintertime temperatures plummeted.

With no definitive answer as to what caused these severe temperature drops, there is room for much speculation. For instance, William Ruddiman, a professor of environmental science at the University of Virginia, argues that the synergies between the Black Death and other miseries inflicted by the Little Ice Age may have had their own effects on the weather. He presented his thesis in the journal *Climate Change* in a 2003 paper entitled "The Anthropogenic Greenhouse Era Began Thousands of Years Ago." Through agriculture and deforestation, he claims, humans became a factor in climate long before the Industrial Revolution. He correlates several of the history's great killing epidemics with drops in CO_2 (as measured in ice cores taken from Antarctica's Taylor Dome). As he explains this relationship, the depopulation following these epidemics led to large-scale abandonment of farmland, which quickly reverted to forest in Europe, China, and North America. During their early years, trees take CO_2 out of the atmosphere, and by Ruddiman's calculations, regrowth of forests following the 1322–51 plagues in China and Europe might largely account for the drop of 10 parts per million in CO_2 measured for that era.

Ruddiman's thesis is controversial. While climate historians such as H. H. Lamb have argued that the cooling and climate gyrations of the Little Ice Age led to disease and then depopulation, Ruddiman explicitly turns this around to assert that plague outbreaks led to depopulation, which in turn caused the cooling. I suspect that he is going to have a hard time finding converts to this bold assertion.

That being said, our power to affect climate today is immensely greater than when the world's population was about 500 million people. And, if Ruddiman is right that humans have already altered climate in the past, we have big problems facing us.

* * *

The previous chapters have all dealt with large-scale events. Even the Little Ice Age, though small compared with the sudden deep freezes and climate swings of the glacial era, is huge compared with anything we have experienced in the modern industrial era. There are other climate cycles, however, that while small, offer detailed insights into how climate impacts human societies, and particularly as to the ways different modern societies have responded to climate once it changes. I am speaking of El Niño.

PART FOUR

El Niño:
The Killer Next Door

15

El Niño: How It Works

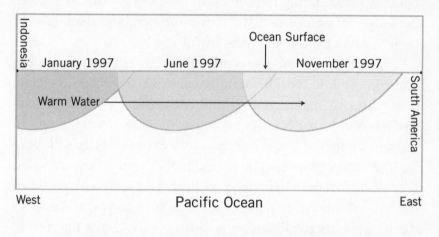

WARM WATER SURGES EASTWARD AS EL NIÑO BEGINS

SINCE THE WANING of the Little Ice Age, the most dominant weather cycle affecting human societies has been El Niño, one of the smallest and most predictable climate upheavals. To put this in perspective, imagine the climate cycles that create ice ages and droughts as waves of various amplitudes rippling back and forth across a pool, sometimes reinforcing and other times canceling each other's effects. Among these ripples, one of the tiniest would be the familiar El Niño. Yet from our pampered perspective at the tail end of the Holocene, El Niño has come to be viewed as a climate monster.

In 1998, a destructive El Niño caused roughly $100 billion

damage around the world, killing thousands through floods in China and Latin America, while droughts in Indonesia contributed to rice riots and political unrest. Cesar Caviedes argues in *El Niño in History* that the effects of El Niños in India and China alone over the past 150 years killed three times more people than the Black Death, and substantially more than the 60 million people who perished during World War II.* In recent years, some scholars have fingered the reverberations of El Niño as contributing to such events as the French Revolution, the failure of Napoleon's Russian campaign, and the failure of Hitler's attempt to reach Moscow in 1941.

That would be quite a record for a climate event that is but a blip compared with the wild swings of the ice ages. Because it is easier to make connections between El Niño and relatively recent historical events such as the disastrous famines of the late Victorian era, it also becomes easier to blame El Niño for the havoc wrought by other, longer-term climate cycles. Just as historians might tend to become overly enamored of the role of disease or deforestation as a historical force, once the focus shifts to climate, it is also easy to see the hand of El Niño everywhere. In the rogues' gallery of climate killers, El Niño may be a mere foot soldier, but because we are repeatedly reminded of its depredations, it looms large in the minds of those who study the impact of climate on history.

If El Niño is one of the smaller of the disruptive climate cycles, it may also be one of the youngest, at least in its present form. From the waning of the last ice age 15,000 years ago until the regular pattern of modern El Niños became established 5,000 years ago, El Niños were intermittent and relatively mild (though it's probable that El Niño–like events happened in the distant past). No one can say for certain what set the stage for modern El Niños, but a number of scientists

*Cesar Caviedes. *El Niño in History: Storming Through the Ages* (University Press of Florida, 2001).

are looking at sea-level rise (between 18,000 B.P. and 5000 B.P. seas rose by about 270 feet as water previously locked up in ice flowed into the oceans) for reasons that become obvious once the workings of El Niño are exposed.

El Niños affect a huge swath of the globe, but the heart of this global weather machine lies in the equatorial Pacific. As with many weather patterns, the basic ingredients of El Niños are the arrangement of oceans and continents, the dynamics of heat and moisture, and the spinning of the globe. At its essence, an El Niño is the result of warm water piling up in the western Pacific. It raises sea level by about a foot in that part of the ocean, with much more dramatic effects beneath the surface as the thermocline (the boundary layer between warm surface water and the cooler waters below) deepens. Then, sloshing back toward the east, the transfer of this heat sets in motion a cascade of oceanic and atmospheric effects that eventually impact the weather from northern Europe to southern Africa and from Australia to Alaska.

The major geographical component of El Niño is the Pacific Ocean. When the Panama land bridge finally pinched off the connection between the tropical Atlantic and Pacific oceans, it also blocked the flow of warm water between the two water masses. Before the land bridge rose, equatorial waters were relatively warm, since the trade winds (recall that the combination of convection where the sun is strongest and the spinning of the globe has the effect of drawing warm air toward the equator and then toward the west) blowing along the equator would shove heated water from the Atlantic into the Pacific. After the Panama land bridge rose, the trade winds continued to blow water westward, but now, instead of drawing water from the Atlantic, the westward-moving water induced currents flowing along the coasts of the Americas, down from California and up from Chile.

As water moves in these currents along the coast, cool

water wells up from the depths to replace the moving surface water. Because of this, under normal conditions the thermocline hovers quite close to the surface in the eastern Pacific. Cool water evaporates less moisture, and the vast amount of moisture that rises from the Amazon basin does not cross the Andes, so the western coast of South America sits in a rain shadow.

The Pacific's vast width—over 11,400 miles at its widest point—provides an enormous expanse of water over which winds can move water and heat can collect from the equatorial sun. The farther west, the deeper the thermocline as the warm water piles up. The combination of equatorial sun and piled-up warm water sets up a powerful engine of rising air near Borneo; indeed, it's one of earth's three great areas of heat distribution. It's also the one equatorial center of heat convection that is not fixed over land. The other two are over the Amazon and Congo rainforests, which also recycle enormous amounts of heat and moisture. According to Richard Fairbanks, a specialist in ancient corals at Lamont-Doherty, the fact that the western Pacific low lies largely over water allows it to move east and west as the earth tilts back and forth during its annual cycle. This may play a role in the El Niño cycle.

This movement of the west Pacific low may be the last piece of the El Niño puzzle to fall into place, since this mobility may be a very recent event. During the peak of the last ice age, sea level was about 270 feet lower because so much water was locked up in ice. As ice melted during the Holocene, sea level steadily rose, reaching its present level about 5,000 years ago. Rising sea level submerged a number of islands, land bridges, and isthmuses (including the land bridge that allowed the peopling of North America). When sea level was at its lowest, Borneo was more than 2.5 times its current size and joined together islands that are now parts of the Philippines and Malaysia as well as Indonesia. As this super island lost 60 percent of its size to rising seas, the low-pressure system became unmoored from the forests, which formerly supplied its heat and moisture, and began to move west to east and east to west.

In all its configurations, the low-pressure system performed the same function: it pushed heat and moisture up into the atmosphere where it could spread north and south. While the low-pressure system distributes heat into the upper atmosphere, the westward warm current below it ultimately hits the continental shelf and splits into two currents, the Japan Current heading north and the East Australian Current heading south. In the North Pacific, this heated water cools and travels eastward across as the North Pacific Drift and then ultimately returns toward the equator as part of the California Current. In our simplified mechanical analogy, the whole thing would look like another giant gear, this time moving clockwise. Mirroring this gear is another system of currents south of the equator. Only this gear moves counterclockwise, since both gears are driving equatorial water westward.

Above the water in the atmosphere, another set of gears is moving as well. This atmospheric circulation is connected to the ocean circulation. For instance, some of the rising hot air in the western Pacific heads north and south, accompanying the warm currents that diverge when the equatorial flow reaches the western edge of the Pacific basin. These flows follow Hadley cell circulation, rising from the lows strung along the equator and then returning at low altitudes. Another mirrored set of atmospheric gears runs east-west following the current; the air rises with the massive convection in the western Pacific and returns at high altitude toward the east and eventually sinks, forming big high-pressure centers in the eastern Pacific, one off the coast of South America and the other off the coast of California. While the Hadley "gear" rotates toward the north at high altitude and then at low altitude back toward the equator, this equatorial gear, called the Walker circulation, has air moving west at low altitude and then returning eastward at high altitude.

So what we have is another vast and intricate set of gears, moving water from the depths in the ocean, circulating surface water north and south of the equator, and circulating air east and west and north and south on both sides of the equator. Driven by the heating of the globe at the equator and the spinning of the globe, these gears distribute heat and rainfall away from the equator.

This climate contraption is a little like a planetary-scale, spring-loaded, wind-up toy. The sun provides the energy; the westward moving water is the spring. At some point, sufficient warm water accumulates in the western Pacific so that the spring is fully loaded. Then something triggers the spring to release and an El Niño is unleashed. Warm water begins spreading eastward, the trade winds stall or reverse, the thermocline deepens in the eastern Pacific, a high-pressure system forms where there was low pressure over Borneo, low pressure forms where there was high pressure off Peru and Chile, and as the configuration of high- and low-pressure systems adjusts around the world, rains fall on Peruvian deserts, storms lash California, droughts afflict the eastern hemisphere from Australia to India, famines afflict the poor, battles turn, and governments topple.

While oceanographers and other climate scientists have gotten pretty good at seeing the signs that signal an impending El Niño, no one can say with certainty what triggers the event, or even whether there is one trigger. Complicating the picture even more is that even though the oceanic and atmospheric gears work together, they work at different speeds. Events race through the atmosphere in a matter of weeks and leave no trace, while the ocean is slow to react and retains an elephantine memory of past events. So—is the adagio of the ocean driving the presto of the atmosphere or vice versa?

The best answer may be "That depends . . ." El Niños don't take place in a vacuum. Until scientists had satellite imagery, it was hard to see linkages between weather systems, which, however, became glaringly obvious when viewed from space. Before the space program, oceanographers and atmospheric scientists had to compare air pressure, sea-surface temperature readings, and changing winds laboriously, and then try to envision the giant weather systems that might connect the dots.

Gilbert Walker, a British meteorologist, did just that and described three oscillations in 1924. Among them was the Southern Oscillation, an atmospheric seesaw that is characterized by swings from high to low air pressure in the Pacific and Indian oceans. When air pressure shifts from high to low in Papeete, Tahiti, and low to high in Darwin, Australia, the trade winds slacken as air spills eastward from high to low pressure. Thirty-seven years later, a pioneering Norwegian-American geophysicist named Jacob Bjerknes, then working at UCLA, connected the air-pressure seesaw in one of these oscillations, the Southern Oscillation, to the sea temperature seesaw in El Niño, envisioning the Walker circulation as the interlocutor between the air and the oceans. Then in the mid-1980s, Mark Cane and Stephen Zebiak, two oceanographers at Lamont-Doherty, constructed a computer model that coupled the two systems successfully enough to predict the El Niño of 1986. Their model was not infallible (it failed to predict some subsequent ENSOs), and it spurred oceanographers to look for other linkages as well.

For instance, El Niños tend to be stronger when the Pacific

Decadal Oscillation (a twenty-year seesaw in air pressure and sea temperatures that affects the higher latitudes of the North Pacific and the more southern waters of the South Pacific) is in its so-called warm phase. Just as the power of a golfer's swing comes from the hips, the background shifts in temperature in the Pacific can mute or enhance the power of an El Niño. The PDO may have gone into its cool phase directly after the powerful 1997–98 El Niño. Perhaps this explains why there has not been a strong El Niño as of 2005 (although a weak event in 2004 produced record snows in the Sierras just around Christmas). The PDO seems to work in concert with the North Atlantic Oscillation, another atmospheric seesaw that shifts every twenty years or so. So far though, scientists have found only ambiguous correlations between the NAO and ENSO.

There are several other climate pulses that might enhance or mute the effects of an El Niño. Again the image of ripples in a pool comes to mind. As these ripples with various amplitudes and frequencies move across the pool, they encounter each other, and every now and then a rogue wave emerges.

One of the biggest questions today is whether El Niños will become bigger and more frequent as greenhouse emissions rise and the globe warms. Observing the frequency of strong El Niños back in the 1990s, some scientists wondered whether El Niño itself was a device for delivering heat in a warming world. The early years of the twenty-first century would contradict that idea, since they have almost all been extraordinarily warm and still El Niño has gone relatively quiescent. It testifies to the complexities of El Niño and its many interconnections to global weather patterns that scientists have produced studies supporting positions on both sides of the question.

It's entirely possible that events that trigger El Niños vary. Scientists have correlated the start of El Niños with everything from volcanic eruptions to sunspot activity. Cesar Caviedes notes that one cycle, the Quasi-Biennial Oscillation (the QBO), is a twenty-six-month seesaw that involves stratospheric winds and thus would be sensitive to solar variation caused by sunspots. In this cycle, high-altitude winds originating in the stratosphere blow parallel to the equator and affect the cooling and warming phases of the ocean as well as set the conditions for the maturation of hurricanes. He cites work suggesting that El Niños tend to occur closer together during periods of low sunspot activity when the QBO seems to have less vigor. In the end though, Caviedes concludes that no single trigger is likely to cause an El Niño, but more likely "only the convergence of several atmospheric, oceanic, and geophysical circumstances makes this happen."

Thus, one of the most predictable climate cycles of all remains unpredictable. Over the long term, ENSOs seem to come on an orderly schedule—11.8 years for major events and 3.5 for minor ENSOs during periods of high solar activity and 7.8 and 3.1 when there are fewer sunspots. According to Caviedes, the variables are such that it remains very difficult to predict the arrival and intensity of any one El Niño. Perhaps this, along with the relative youth of the present day El Niño cycle, helps explain why strong El Niños cause massive die-offs of sea birds, seals, and other animals. It takes time and regularity in nature for animals to develop an adaptation.

Societies have contended with El Niños since civilization first established a toehold in Asia Minor. In the West at least, all weather was perceived as local until the mid-nineteenth century. Early astronomers and meteorologists had neither the tools nor the perspective to predict the rhythms of this global phenomenon, or even know that a global weather phenomenon was taking place. And so, time and again, El Niño has us.

16

El Niño Meets Empire

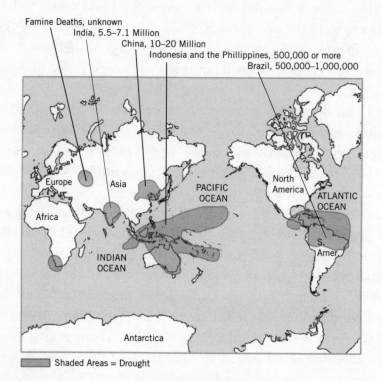

Famine Deaths, unknown
India, 5.5–7.1 Million
China, 10–20 Million
Indonesia and the Phillippines, 500,000 or more
Brazil, 500,000–1,000,000

Europe
Asia
PACIFIC OCEAN
North America
ATLANTIC OCEAN
Africa
INDIAN OCEAN
S. Amer.
Antarctica

Shaded Areas = Drought

GLOBAL DROUGHT, 1876–78

CLIMATE'S CAPACITY to inflict misery rises steeply when imperial arrogance and ideology hinder a society's adjustments to extreme weather. In fact, some students of El Niño argue that the fatal chemistry of colonial arrogance and powerful El Niños shattered indigenous cultures during Victorian times and played a crucial role in the creation of the Third World.

Historian Mike Davis makes a strong case in his extraordinary book *Late Victorian Holocausts: El Niño Famines and the Making of the Third World.* * He explores the repercussions of nineteenth-century famines that accompanied droughts in India, parts of China, Indonesia, East and southern Africa, and northeast Brazil in lockstep with ENSO events. His meticulous accounts of the scale of suffering are all the more horrifying for the dispassionate way he lays out his case, and the way in which Queen Victoria's emissaries amplified the casualties through their beetle-browed adherence to "free market" economics. Davis writes: "Millions died, not outside the 'modern world system,' but in the very process of being forcibly integrated into its economic and political structures. They died in the golden age of Liberal Capitalism; indeed, many were murdered, as we shall see, by the theological application of the sacred principles of Smith, Bentham and Mill."

The interplay of economic ideas, colonial power, and weather led to famines in India and China that the great naturalist Alfred Russel Wallace cited as one of the "most terrible failures of the century." Davis describes these famines as the missing pages of the Victorian era that falsify the conventional wisdom celebrating the emergence of the modern market economy. Even at the height of the famines of 1877, 1878, and the 1890s, India continued to export grain to England. (In *El Niño—History and Crisis*, Richard Grove and John Chappell point out that Europe was then in the great agricultural depression, perhaps itself related to the powerful El Niños of those years.†)

---

*Mike Davis. *Late Victorian Holocausts: El Niño, Famines, and the Making of the Third World* (Verso, 2001).
†Richard Grove and John Chappell. *El Niño—History and Crisis* (White Horse Press, 1999).

Davis has an agenda, to be sure. As he puts it, "a key thesis of this book is that what we today call the 'Third World' . . . is the outgrowth of income and wealth inequalities—the famous 'development gap'—that we shaped most decisively in the last quarter of the nineteenth century, when the great non-European peasantries were initially integrated into the world economy. . . . By the end of Victoria's reign . . . the inequality of nations was as profound as the inequality of classes."

While much has been written about the events and the evolution of world market forces, Davis's contribution is to integrate the nineteenth-century famines with the El Niño cycle, whose links to monsoon rain failures have come to light in recent decades. The two great events of the period in question were the El Niños of 1877–78 (one of the most powerful ever recorded) and 1899–1900. For much of Asia, the key variable in the ENSO cycle is the monsoon. The word comes from the Arabic word for season and refers to the fair winds that trading ships relied upon. The wet part of the cycle occurs during the summer, when winds, drawn over the ocean and then northward, bring up to 90 percent of the annual rainfall in some parts of India between June and September.

As with any climate phenomenon, the monsoon is vulnerable to a bewildering variety of factors, but some are more important than others. One of these is the difference in temperatures between the Eurasian landmass and the oceans to the south. As winter gives way to spring, snow cover melts over the continent, particularly on the vast Tibetan plateau, which, though relatively far south, is so high (averaging about 15,000 feet) that it gets a good deal of snow cover in the winter as well as intense solar heating as the sun moves higher in the spring sky. This sets the stage for the onset of the moist summer monsoon and the rains that follow.

While snow reflects heat back into space, once exposed, the darker soil absorbs heat. Consequently, the plateau as well as the surrounding continent shifts from being an exporter of cold dry air in the winter to an importer of warm winds from the south in summer. Temperatures over land warm much more rapidly than ocean temperatures. The difference in air temperatures over the relatively cool waters and the hot land sets up a vigorous circulation as convection over the plateau sucks moist air north and west over India. Because of this, the duration of the snow cover on the plateau and the interior of the continent has a lot to do with the strength or weakness of the monsoon.

When things are working right, the first monsoon delivers rain in southeast India around June 1, and within two weeks the rains have spread north and west to water the entire subcontinent. The winds surrender whatever moisture they still retain as they are drawn upward over the Himalayas, making the towns and cities in the Indian foothills some of the wettest places on earth. As a kid, I remember trying to imagine what it must be like to live in Cherrapunji, India, where up to 39 feet of rain have fallen in a year, roughly ten times the amount we were getting in New York State.

Naturally, things do not always work perfectly to deliver a good soaking. The longer the snow lasts, for instance, the more likely it will be that the subsequent monsoon is weak. Or a high-pressure system that typically sets up over the high mountains in northern India might decide to linger into summer, blocking water-bearing storms from pushing north. Or the Indian Ocean warms up, reducing the temperature gradient with the Eurasian interior and diminishing the vigor of the monsoon.

Or, an El Niño arrives and the monsoon fails altogether. This is not a certainty (the two major El Niños in 1982–83 and 1997–99 were not accompanied by drought in India for reasons that will be discussed later), but from 1789 until 1922 (when El Niño went into a two-decade quiet period), every

El Niño was accompanied by varying degrees of drought in India. During an ENSO, the shifting geometry of the Walker Circulation—the atmospheric pressure systems strung along the equator—can temporarily strand India under high pressure where dry air descends. The failure of the Indian monsoon tends to precede other El Niño signals, such as the warming of the waters off Paita on the coast of Peru.

The first hints of what was to come for India during the great El Niño of 1877–78 came in 1876 when a meager monsoon left parts of Madras with a quarter of their normal rainfall. Indeed, drought afflicted India several times during the 1870s, even in the absence of an El Niño, suggesting that the monsoon was in one of its multidecade weak phases, periods that tend to amplify the drying effects of El Niño on India.

Those counting on spring rains were again disappointed in April and May, when again rainfall was but a fraction of normal. June, the customary time of arrival for the monsoon, came and went with virtually no rainfall. So did July and August. Roughly half of India's cultivable land received no rainfall at all during a period that ordinarily delivers 80 percent of the annual rainfall. When the rains did finally arrive in September, it was too late, and in fact the resulting deluge spawned a raft of miseries.

Those are the bare bones of the 1877–78 drought. For over 100 million Indians, the dearth precipitated unimaginable horrors. Meticulous bureaucrats, the British documented the decline in crop yields, though they were less punctilious in tabulating the human tolls. (There is some irony in the fact that Davis's most devastating indictments are built upon data unselfconsciously collected and published by the Raj.) In the famine districts in Madras for instance, the percentage of crop losses ranged between 66 percent at best and nearly 95 percent in some districts.

Without crops, and without adequate relief, people began

to starve by the millions. They ate shoe leather, then livestock, then dogs, then plough animals, a desperate move since the action left the farmer with no ability to plough the land for the next planting. As the hunger deepened, people on occasion began to eat each other, and when they became too weak to move, dogs and wild animals ate them. Here's a sampling of one of the many firsthand accounts cited by Davis, this from a "relief" camp: "The dead and dying were lying about on all sides, cholera patients rolling about in the midst of persons free of the disease . . . for shelter some had crawled to the graves of an adjoining cemetery . . . the crows hovering over bodies that still had a spark of life in them. . . ."

The relief camps were but one way in which the British amplified the effects of the drought. These camps were the creation of Sir Richard Temple, described by Davis as Viceroy Lytton's "enforcer." Lord Lytton assumed the duties of viceroy in 1875, just a year before India began its descent into drought. Edward Robert Bulwer-Lytton was the son of Edward Bulwer-Lytton, the novelist who wrote the words "It was a dark and stormy night," now immortalized by the annual Bulwer-Lytton contest for bad writing. The son carried on the family tradition. A poet who wrote under the name of Owen Meridith, Viceroy Lytton was dogged by accusations of plagiarism, but he had one notable fan, Queen Victoria. According to Davis, the viceroy's attitudes toward humanitarian relief were that it was "cheap sentiment," and he was steeped in the free-market ideology of Adam Smith who, in *The Wealth of Nations,* had written about an earlier Indian drought, "Famine has never arisen from any other cause but the violence of government attempting, by improper means, to remedy the inconvenience of dearth." Also in the air were the teachings of Thomas Malthus dusted with a smattering of social Darwinism. Davis quotes Lytton as stating that population "has a tendency to increase more rapidly than the food it raises from the soil." Lytton gave an early indication of his insensitivity to the suffering of his subjects when, as the meager rains of the summer of 1876 caused Indians and officials alike to worry about the prospects of disaster, he sponsored a weeklong feast for 68,000 officials to celebrate Victoria's assumption of the title of empress of India.

Into this heady mix of imperial power and intellectual ferment stepped Richard Temple, who had been castigated for a relief effort a few years earlier when he'd authorized the import of grain to stave off famine and, clearly being out of step with the program of other Victorian officials, dispensed a ration that would actually sustain life. Now, in 1877, he was hell-bent on showing that he too was a true believer in the omnipotence of a free market unencumbered by doles and safety nets.

As the drought gathered intensity, Temple decided to criminalize the urge to help one's fellow man, by making it illegal in the state of Madras to give relief donations that might interfere with the price of grain on the markets. His contribution to history, however, was the Temple wage, a supposedly scientifically derived ration that would sustain men doing heavy labor. That wage, one pound of rice per day, was half what felons received in Indian prisons. Unsupplemented by meat or vegetables, it provided 1,627 calories a day, which, Davis offers, is only 127 calories more than an adult consumes each day while in a coma, 123 calories *less* than the starvation rations of the Nazi death camp Buchenwald and more than 2,200 calories less than the approved diet for an Indian male doing heavy labor today. The inmates of these camps were not only supposed to survive on these rations, but also build railroads and canals. The results were predictable. Grown men shrank to 60 pounds. By one calculation, the deaths each month in the camps annualized to a rate of 94 percent.

Things were little better outside the camps. The areas

served by railroads and granaries, areas integrated into the free market, fared worse than the boondocks. Perhaps part of this discrepancy can be accounted for by the fact that, at the height of the famine, grain exports surged. In 1877, a poor crop in England raised prices, and India exported double the amount of wheat in 1877 that it had in 1876. The government also refused to suspend ruinous taxes. Corruption and greed also played a role. Grove and Chappell, citing a study of the period, argue that "the famine was aggravated by grain dealers who used the new railway system to concentrate grain in areas where it fetched the highest prices, so that those least able to afford it starved even more rapidly than in pre-railway droughts."

The British government estimated that 5.5 million died, but neglected to include in their calculation entire states that suffered through the droughts. Other estimates range upward from seven million.

India was but one of the nations hit by famine that year. Great swaths of China suffered drought and crop failures during the El Niño. Desperate parents sold, and sometimes killed and ate, their children. Chinese deaths added 10-to-20 million to the toll, with some prefectures suffering 95 percent mortality. Millions more died in Brazil, Indonesia, Africa, and other places affected by the most powerful El Niño in 500 years.

The horrors of these famines got back to England thanks to the reporting of crusading journalists like William Digby and Robert Knight, letters to the London *Times* of Florence Nightingale, and others sent back by missionaries and travelers. Many English both in India and (once made aware) at home were roused by the horrible reports and tried to help with contributions. Moreover, the disaster occasioned a good deal of introspection, and a royal commission, charged to leave no stone unturned. One would think then that the next time drought threatened famine in India during Victoria's reign the government would be prepared. One would be wrong.

Perhaps the British overlords thought that drought was a thing of the past. Even those not convinced that a replay of 1877 was out of the question could take some solace from famine codes and famine funds instituted in the intervening years. After laying low for most of the 1880s, El Niño returned with a vengeance toward the end of the nineteenth century. In quick order, an El Niño in 1896–97 shifted to La Niña (the cold event that brings its own troubles), and then back to an extremely powerful El Niño to close out the century in 1898–99.

When the rains failed in 1896, most of these preparations were exposed as useless. Much of the famine fund, for instance, had been diverted to help fund a war effort in Afghanistan. History repeated itself, as Mark Twain would have it, in rhyme rather than exact replay. True to the obtuse style of his predecessor, the viceroy, Lord Elgin, traveled through the country and saw relative prosperity in the Central Provinces, even though, Davis asserts, there were reports of 10 percent monthly mortality in the region. Once again, India's rulers believed that modern markets had eliminated the possibility of famine. Elgin said as much in a letter to Queen Victoria. Once again, the elite looked upon the famished as layabouts. Once again, the rulers took refuge in Malthus and social Darwinism. Davis quotes Sir John Strachey as taking "hope and encouragement" from the fact that the famines of the 1890s killed the poorest of the poor.

By the time the rains returned with the La Niña of 1898, probably (estimates vary) more than 10 million Indian poor had died. Thinking the worst was over, the press wrote about the event in the past tense, calling it "the famine of the century." Ordinarily, this would be a safe bet, since only two years remained until 1900. But if 1877 was the driest year in

Indian history as far as could be known, 1899 ranked second. While the 1877 drought desiccated half of India's land, the premillennial drought left more than two-thirds unusable.

Now it was Lord Curzon's time to play the role of dogmatic viceroy so ably acted out by Lytton and Elgin. The British public was distracted by the dramas of the Boer War and the Boxer Rebellion (itself to some degree a product of El Niño famine in China). With only a muted outcry at home, Curzon was able to reduce rations for the poor and declare millions ineligible for any relief at all. The overseers of relief camps set their workload targets according to the standard of a well-fed adult male doing heavy labor, according to Davis, and then cut rations proportionate to the degree the workers fell short. In one sense, Curzon went Lytton and the Nazi masters of Buchenwald one better, since the minimum allotment was actually less than the Temple wage that had starved so many people twenty years earlier. Once again, speculators exported grains to England from areas still blessed with rain as millions starved. The body count for 1899–1900 ranges from 3 million to 10 million.

As was the case in 1877–78, India's misery was replicated throughout the El Niño drought belt in China, East and southern Africa, Brazil, Indonesia, and the Philippines. Estimates of the death toll from the El Niños of the 1870s and 1890s are about 50 million people. Partly as a result, population growth in many parts of the tropical world stagnated and even dropped in districts of India and China between 1870 and 1910. Countless millions who survived the famines lost their lands, their livelihoods, and their traditional way of life.

Davis does not argue that ENSO by itself changed history. He asserts that famines are not the result of inadequate food production but of inadequate food availability. "I argue that *ecological poverty*—defined as the depletion or loss of entitlement to the natural resource base of traditional agriculture," he writes, "constituted a causal triangle with increasing *household poverty* and *state decapacitation* [by which he means the inability of the government to deliver infrastructure that might ameliorate the effects of drought] in explaining both the emergence of a 'third world' and its vulnerability to extreme climate events." He cites evidence that prior to British interference in China and India, both societies had proved capable of dealing with droughts without widespread famine (although Grove cites accounts of appalling starvation in India during the El Niños of 1596 and 1630). Running through the entire book is an indictment of the free market for the failure of its invisible hand to prevent mass starvation.

First of all, it should be said that Davis performs an enormous service by simply offering the details of these great famines and their links to El Niño. It is all too easy to read about the modernization of India without any mention at all of the tens of millions who died as the British forcibly integrated India into the world economy, or, to be more precise, into their empire. The *Columbia History of the World,* for instance, recounts the history of British control without ever mentioning the famines and credits the integration into the market economy for unifying the country.*

Davis's most powerful evidence of the perfidious nature of Albion is his voluminous evidence of British self-dealing. Officials used tariffs and other means for the benefit of English trade; they used taxes during times of famine to confiscate land; they used their preferential status to dump inferior products into the captive Indian market to sustain otherwise uncompetitive industries in England. In other words, Davis makes a good case that what the average Indian experienced was anything but a free market. In reality, the poor Indian had no choice how to deploy his labor, nor any recourse.

Columbia History of the World (Columbia University Press, 1990).

El Niño went quiet for over fifty years starting in 1926, with just a couple of strong events during the period. A strong El Niño in 1972 produced drought throughout India and affected 50-to-100 million people, but produced only 800 deaths from famine. Then, when El Niño regained its punch in 1982, for some reason it uncoupled from its prior correlation with drought in India. The ENSOs of 1982 and 1997 delayed the onset of the monsoon somewhat, but did not produce drought on the scale of previous El Niños of similar magnitude. This may have something to do with changes in the cycles that affect the background state of El Niños, or perhaps the general warming of the globe during the past two decades. The setup of pressure systems has shifted to the south and east during ENSOs, moving away from India one impediment to the monsoon—the high-pressure system over the Indian Ocean that in previous El Niños blocked the northward flow of moist winds. It's possible too that a warmer Eurasian land mass is providing more early energy for the monsoon. The point is that India's capacity to deal with extreme drought has not really been tested in over a century, and now with over a billion people and with ever increasing competing claims on the world grain market, the Indian government can no more afford to be complacent about their ability to deal with the next megadrought than could the British during the Raj.

How might the world be different if the fifty-year lacuna in El Niños had taken place in the last half of the nineteenth century rather than in the middle of the twentieth. A stretch of fifty years with relatively normal monsoons would have seen a dramatic increase in Indian population, and would have deprived the British of a crucial weapon in their effort to expropriate land and create a labor force (the government and private interests seized enormous amounts of land during both El Niños, and those forced to marginal lands during the 1870s were more vulnerable to the droughts of the 1890s). For each person killed outright by famine or disease, many more were ruined. Without this ecological poverty (to use Davis's term), it's possible that Indian resistance to the enormous transfer of wealth to England (at one point in the early twentieth century, India accounted for 73 percent of the U.K.'s international trade credits) and to the destruction of traditional agriculture might have been more effective.

The El Niños provided similar opportunities to the British in China and Africa, the Dutch in Indonesia, and the landowning elites in many other countries. El Niño did not create the Third World, but it provided a powerful instrument to smash the traditional economic and social structures that formerly promised self-sufficiency in many parts of the tropics.

A few historians argue that on occasion El Niño has changed history by emboldening the poor to rise up against an oppressive government. It's widely accepted that the bread riots of 1788 were the prelude to the French Revolution. Richard Grove of the Institute of Advanced Studies at the Australian National University places these riots in the context of the severe El Niño of 1789 to 1793. In a letter to the scientific journal *Nature,* he asserted that the cold winters in Europe in the preceding two years were linked to the subsequent El Niño (though it's also possible that they represented a reverberation of the Little Ice Age which still had Europe in its grip). Elsewhere, Grove has argued that these riots and earlier food riots under Louis XIV's reign opened the eyes of the poor to the possibility of civil disobedience.

Few historians have come forward to embrace this thesis (Davis is one who dismisses it). In his letter to *Nature,* Grove points out one interesting sidelight of this El Niño. He notes that Alexander Beatson, the governor of St. Helena in 1791, suggested that the droughts that year in India (which killed 600,000 in Madras alone), Montserrat in the Caribbean, and St. Helena were all part of a connected phenomenon. He made this suggestion more than 130 years before Gilbert Walker advanced his theories about the Southern Oscillation.

Serious scientists and historians are careful to distinguish between the fact that some event might have taken place against the backdrop of a famine, and the role that famine may have played in altering the course of history. While historians surely will continue to argue with Davis about the role of El Niño in the creation of the Third World, or in altering the fate of any of the multitude of societies that suffered its impacts, the synchronicity of all these crises around the world bears out the larger point that climate anomalies correlate with historical inflection points. Just as it is compelling to realize that human ancestors underwent rapid evolutionary change at a point when many other species were changing as well in the face of climate challenges, it is arresting to discover rapid social, economic, and political change in far-flung corners of the world linked by no common event other than the ripples of El Niño.

Some scholars argue that these repercussions can take place very far afield from El Niño's Pacific domain. Every schoolchild learns at some point that the Russian winter defeated both Napoleon and Hitler. Cesar Caviedes argues that it was not so much the Russian winter that proved the graveyard of the French army in 1812 and German divisions in 1942–43, but a wild series of storms, cold fronts, and thaws that typify central Asian weather as an El Niño winds down and shifts to a La Niña. The Russian winter is rarely a picnic, but Caviedes, with backing from some German meteorological research, argues that the combination of mud and storms took the blitz out of both Nazi German and Napoleonic armies.

At first blush, Caviedes's argument seems a stretch. For one thing, the degree to which El Niño affects the Russian winter is open to question. The proximate climate cycle in the North Atlantic is the NAO, and, as noted, the links between NAO and ENSO are still not fully understood. Moreover, the El Niño in question occurred during 1940–41, not during the winter of 1942–43, when the Russians trapped the German Sixth Army in a pincer as General Paulus failed in his attempts to take control of Stalingrad. Of course, many other large factors affected the fortunes of Paulus as well, including attenuated supply lines that would be difficult to defend in any weather, the fierceness of the Russian resistance, underscored by Stalin's fanatical concern that his namesake city not fall into German hands. Also, the Soviets simply outmaneuvered the German commanders.

Here again though, weather gave some crucial advantages to the home team. Incessant storms often prevented aerial resupply and bogged down Panzer divisions in a late November rescue attempt launched by Field Marshal Erich von Manstein. As fall turned into winter, mud gave way to sleet, snow, short thaws, and renewed storms. While this sounds like a normal winter succession, there are significant differences with prevailing winter patterns. Caviedes cites the work of a German meteorologist, Klaus Fraedrich, who studied over a century of observations to analyze the impacts of ENSO. Fraedrich argues that a southward shift of the cyclonic track that correlated with El Niños during that period led to greater fluctuations in weather. Supporting this interpretation, Caviedes notes that, during the subsequent La Niña, the following winter was more stable, even if it was cold, and that the surviving Germans not only managed a deliberate withdrawal, but also thwarted a Russian offensive.

Caviedes is on even less certain ground when he ascribes a role to El Niño in the catastrophic failure of Napoleon's campaign to capture Moscow in 1812. While there was an El Niño at that time, historians don't have detailed accounts of the daily weather. This is not to say that unusual weather is irrelevant. Bonaparte had the bad luck to mount his campaign during a period of general cooling that began around 1810. This was one of the last paroxysms of the Little Ice Age, perhaps enhanced by the chilling effects of volcanic eruptions, which throw sun-blocking volcanic ash into the atmosphere. Earlier that year, a major eruption of Soufrière on the island of St. Vincent would have had a bigger impact on global weather than did the eruption of Mount Pinotubo in the Philippines in 1991, which temporarily reversed the warming trend of the 1980s and 1990s. Unusual weather may well have aggravated Napoleon's troubles in Russia, and changeable, extreme weather did impede the movement of his army. But Caviedes may see the fingerprints of El Niño when another climate culprit is at work.

Scientists have offered linkages between El Niño and many notable historic events, including the sinking of the *Titanic* (based on linkages between southward surges of icebergs and ENSO) and the potato famine that killed at least 1 million people in Ireland. (The continual rainfall and unusual warmth typical of El Niño years in Ireland created the perfect conditions for the spread of *Phytopthora infestans,* the fungus that ruined the crop.) The sinking of the Spanish Armada took place during anomalous storms of the El Niño of 1588, and offers a case of climate Rashomon, with some scientists focusing on El Niño while others see the fingerprints of Arctic-driving NAO events. While there may be merit in some or all of these assertions, they mostly bear out the notion that once one acknowledges the fingerprints of ENSO on one historical event, one sees them everywhere.

For instance, some scientists associate the Dust Bowl in the United States in the 1930s to the impact of the 1932 El Niño, while others offer it as proof of the potentially destructive role of La Niña. Moreover, between them, El Niño and La Niña can chew up a lot of years, meaning that at times when El Niños are frequent and active it is a statistical probability that a certain number of historical events will take place under one or the other's influence.

To return to the analogy of the serial killer, the examples of El Niño–caused starvation in India, China, Brazil and elsewhere offer strong evidence that climate change can play a crucial role in historical change. As is the case, however, with many serial murderers, the killer has been invited into the house. The British rulers magnified the effects of El Niño. Time and again in centuries past, societies have set the stage for calamity by loading the gun that the killer will use.

17

A Taste of Things to Come?

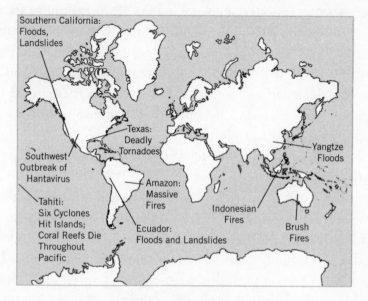

Southern California:
Floods,
Landslides

Texas:
Deadly
Tornadoes

Southwest
Outbreak of
Hantavirus

Tahiti:
Six Cyclones
Hit Islands;
Coral Reefs Die
Throughout
Pacific

Amazon:
Massive
Fires

Ecuador:
Floods and Landslides

Indonesian
Fires

Yangtze
Floods

Brush
Fires

1997–98: THE $100 BILLION EL NIÑO

ONE QUESTION LOOMS over the role of climate in history: are we getting better at adapting to change? El Niño offers a useful gauge, because climate specialists have a pretty good idea of the comparative strength of El Niños stretching back thousands of years. While modern civilization has not been tested by the extreme climate shifts that characterized the glacial era, just a few years ago the world endured one of the strongest El Niños ever measured.

The monster El Niño of 1997–98 offers a case study of the interconnections between climate, economic activity, and politics. Perhaps its most notable casualty was the end of Indonesia's Suharto regime on May 21, 1998. The Asian financial crisis of 1997 contributed to Suharto's fall, and the origins of that meltdown were in the international currency markets. Moreover, after thirty-two years of corruption, cronyism, and nepotism (nicknamed KKN by Indonesians for the first letters of the words), Indonesians were fed up. If the fallout of the regional financial crisis and misrule (along with criminally misguided agricultural and forestry policies) built the pyre, El Niño provided the spark. A two-year drought scorched Indonesia's rice crop and unleashed fires that burned an area as large as Costa Rico and sickened tens of millions with its accompanying pall of smoke.

The El Niño was the strongest of the twentieth century and one of the most powerful ever recorded (the 1982–83 El Niño was more intense in some ways, but much shorter in duration). According to the National Oceanic and Atmospheric Administration (NOAA), which collects and analyzes global weather data, the El Niño raised air temperatures over land by about .8 degrees centigrade or 1.4 degrees Fahrenheit. (This is roughly equivalent in magnitude to the global warming of the past century.) The next-largest El Niño of the century took place in 1982–83 and also raised havoc. Also, according to NOAA, global temperatures set a record for warmth during every month from January through June in 1998.

In Indonesia, the costs of suffocating smoke and a shriveled rice crop contributed to rampant inflation even as the economy contracted. During the first half of the year, prices soared at an annualized rate of 80 to 85 percent. In the first quarter of 1998, the Indonesian economy contracted by 8.51 percent. By May 1998, the rupiah, the Indonesian currency, had lost 79 percent of its value compared with exchange rates the previous July. (This compares with 30 to 40 percent declines for other Southeast Asian currencies during the so-called Asian meltdown.)

For the average Indonesian, this was disastrous. Per capita income fell by 75 percent, from $1,200 a year to $300. Actual, as opposed to reported, unemployment was estimated to be 40 percent. In May 1998 alone, the price of food rose by 4 percent, but for many Indonesians food was scarcely available at all after two years of failed rice crops.

Most poor Indonesians depend on rice as the center of their diet, and, as is the case of most countries where rice is a staple crop, most rice is grown locally. Rice does not command the price that would justify a robust export market, and with a couple of exceptions, Indonesia imported negligible amounts of the crop until 1998. During the fall of 1997, the monsoon arrived about forty-five days late, in mid-November. When the rains came, however, they were meager. Out of thirty-three weather stations scattered through Indonesia, thirteen reported the lowest rainfall on record. After shrinking by 3.3 percent in 1997, the rice crop shrank by another 6.25 percent in 1998. The government was forced to enter the market massively, importing nearly 6 million tons of rice. For that year, Indonesia was the largest rice importer on the planet.

Despite these imports, the price of rice rose, nearly tripling from its levels in March of 1997. Catastrophic for the poor, this increase might have been good for farmers. Unfortunately, the devalued rupiah had the effect of increasing the price of fertilizers and pesticides fivefold. In some of the outlying provinces, such as Irian Jaya, people began to starve to death.

The fires that accompanied the drought imposed their own hardships. Across the region, an estimated 70 million people suffered respiratory ailments. It's possible that the fires exacerbated the drought, since the particles that make up smoke are so fine that they can offer too many nuclei for a given amount of water vapor. The result, according to various studies of fire and rainfall in tropical forests is that raindrops never get heavy enough to fall.

The country careered toward anarchy. Before the currency crisis and El Niño, an estimated 22.5 million Indonesians lived below the poverty line. By the spring of 1998, this number had risen to more than 100 million people, roughly half the population. In the cities, mobs launched periodic raids on ethnic Chinese. In the countryside, vigilantes sometimes lynched so-called "ninjas" in a bizarre manifestation of mass hysteria in which martial arts movie imagery became overlaid onto traditional myths. The unfortunate victims often turned out to be innocent outsiders in the wrong place at the wrong time. Then in the spring of 1998, students and others organized massive protests. Public discontent only increased when the desperate government acceded to demands of the International Monetary Fund and removed price controls from fuel and electricity in return for $3.4 billion in new loans.

Mohamed Suharto relinquished power on May 21, 1998. Six years later, rice production remained below 1996 (pre El Niño) levels, and while the nation does not need to import as much rice as it did during the El Niño, it is much farther from self-sufficiency than it was before. In 2004, 35 million Indonesians lived below the poverty line, according to the World Bank, a level nearly 50 percent higher than before the El Niño. Moreover, according to the bank, another 70 million still must get by on less than $2 a day. With Sumatra reeling from the effects of the 2004 tsunami, these numbers will surely increase.

Corruption continues, despite some valiant efforts at reform, and in some respects has worsened. Illegal logging, rampant under Suharto, has actually worsened in the years since. Some of Indonesia's islands, including Sumatra, will probably lose all their ancient forests in just a few years, with catastrophic consequences for both wildlife and people. The fires during past El Niños were exacerbated by ill-conceived agricultural projects, which exposed highly flammable peat soils. Deforestation has also reduced regional rainfall. The next strong El Niño will likely bring a catastrophe that will dwarf the events of 1998.

Is it reasonable to attribute the fall of Suharto to El Niño? The currency crisis had nothing to do with the weather, and it could be argued also that the scale of Suharto's corruption guaranteed a crisis at some point. In 2004, the nonprofit business group Transparency International ranked Suharto as the most corrupt head of state of the prior twenty years, estimating that he and his family stole $15-to-$35 billion from the country. That's quite a distinction, given that Suharto's peers include thieves such as Mobutu of Zaire and Ferdinand Marcos of the Philippines.

Isn't it possible then that almost anything might have brought this teetering regime down after thirty-two years in power? Around the world, the forces of markets and democracy were eroding the power of kleptocrats. Would not those forces have brought down Suharto by themselves? Moreover, other nations survived the same El Niño relatively intact. The neighboring Philippines was hit by drought as well. At one point, 1 million Filipinos faced famine, but the nation did not descend into anarchy.

The simplest way to put El Niño into perspective is to imagine how 1997 and 1998 might have played out in Indonesia had the weather remained normal. The country would still have had to contend with the currency crisis, but the poor and middle class would not have been faced with food shortages and punishing price increases for rice. Nor would productivity and health have suffered from the effects of the fires. It's possible that without the food and health crisis, Surharto might have survived, but, in retrospect, it also seems clear that the effects of the El Niño ensured that his regime would not survive. Just as failure to deliver rains and the crops they nourish brought down ancient pharaohs and Mayan rulers, it helped bring about the fall of a modern dictator.

Haile Selassie represents another instance where ENSO helped hustle a twentieth-century ruler off the stage. Regularly recurring famines resulting from drought have killed millions of Ethiopians down through the centuries. Many climate influences buffet Ethiopian weather and the country has a number of microclimates, but drought correlates strongly to ENSO events, most often starting at the tail end of an El Niño and continuing during the following year. During an El Niño, a complicated series of climate gears (the El Niño warms the upper atmosphere throughout the tropics, and with warm air piled upon warm air, there is less vertical movement of evaporated moisture) shuts down the big rains called Kremt that normally water the country from June to September, and then delivers rain during what is ordinarily the dry season from October to January. Just as the country is recovering from El Niño, the "cold" event, or La Niña, that normally follows an El Niño delays the northward movement of the rain belt that ordinarily delivers rain between February and May.

The result is that crops and livestock are starved for moisture when they need it most and then drowned during the harvest season. The puny crops are vulnerable to fungal blights and pests in the abnormally wet harvest season, seeds germinate before they are harvested, and weakened livestock die from disease. As crops and animals die, people soon follow. The powerful El Niño of 1888 produced the Great Ethiopian Famine that killed one-third of the population and 90 percent of the country's livestock. The El Niño of 1972–73 contributed to the 1974 famine that killed 200,000 people in the northern region of Wollo. And the intense El Niño of 1982–83 set in motion the drought of 1983 to 1985 that affected 7.5 million people and killed 300,000.

Like Suharto, Haile Selassie was ripe for a fall. He had been in power since 1941 and was increasingly out of touch during the final years of his reign. He made the mistake of accepting advisors from the Soviet Union, many of whom went busily to work organizing Marxist cells and encouraging discontent. When I spent some time in Addis Ababa toward the end of the regime, the ancient and beguiling city was awash with rumors, and the emperor, isolated in his compound, was an all-but-irrelevant figurehead. His efforts to deal with the famine proved ineffectual.

I doubt that that there was any real fervor for the communist Dergue beyond the ranks of the intellectuals, but the public was fed up with corruption and ineptitude, and if most people did not welcome the Dergue takeover, they were at least willing to see whether these austere ideologues could deliver something better. They didn't, of course, and because of their own brutality and ineptitude, they too gradually lost legitimacy.

Weather, having no ideological bent, helped dump the Marxist regime as well. The famines following the 1982–83 El Niño exposed the helplessness of the Mengistu regime, and if that did not get the point across, famines following the 1986–87 El Niño did. Mengistu lasted through two more droughts (neither related to an El Niño), and then lost power in the middle of the El Niño–related famine of 1991.

By 1991, the regime could no longer count on assistance from the former Soviet Union, and for many political scientists that alone might provide necessary and sufficient conditions for the fall of an unpopular regime. Climate events do not occur in a vacuum, and climate stress focuses crisis and discontent. Moreover, if people trust the intentions of a government, they are less likely to take to the streets when an El Niño puts a nation to the test.

This was underscored by the effects on Ethiopia of the 1997–98 El Niño, the same event that contributed to the fall of Suharto. Once again, drought and unseasonable rain hit the country, reducing harvests in these years by as much as 40 percent. This time, however, the Ethiopian government was more proactive. Using models based on sea-surface temperatures and other anomalies, the government issued a warning that the summer rains would be diminished and erratic on May 29, 1997. In a case study of the impacts and responses to this El Niño, Tsegay Wolde-Georgis of the Ethiopian embassy in Washington, wrote:

> [T]he Ministry of Agriculture recommended the planting of drought-resistant crops, such as chickpea, and the replanting of failed crops with fast-maturing ones, such as teff and lentils. It also recommended the provision of seeds to farmers until the end of August and the protection of crops through the free distribution of pesticides. In addition, it advised the use of all newly constructed micro-dams and ponds by farmers, as well as the building of irrigation canals. Farmers were also advised to plant potatoes and convert lost crops to feed the animals. The mass media was to be used to educate the people on the response actions. . . .
>
> As the season progressed, the Ministry of Agriculture advised farmers to replant their crops especially when the rains returned to some areas in August. The zone officers provided the seeds on credit in order to take advantage of the rains. The October–November 1997 floods disrupted the normal harvest of crops, and local governments or-

ganized popular campaigns to help farmers gather their crops. . . . on Nov. 12, 1997, the NMSA urged farmers to "gather their harvest before an untimely rain expected over the various parts of the country."

Ethiopians suffered terribly during this El Niño, but largely escaped the mass starvation that followed powerful El Niños in the past. The same holds true for many other countries around the world: they endured suffering and economic harm, but did not descend into chaos. Somehow the global economy managed to ride out the most powerful El Niño of 125 years.

Such resilience may provide an important data point in addressing the question of whether modern economies are better equipped to withstand changes in the weather than our forbears. The results would appear to be a score for the modern market economy, since less powerful El Niños in the past have had outsized effects, while more recent El Niños have not produced carnage on an equivalent scale.

An El Niño, as we have observed, is not on the scale of a Little Ice Age or the A.D. 536 event. Nor is it of the same order of magnitude of the climate tremors that swept around the globe 8,200, 5,200, or 4,200 years ago. We simply don't know whether this resilience, which has somewhat muted the impacts of recent El Niños, is sufficient to ameliorate the effects of the longer, more severe droughts, floods, storms, and shifts in climatic zones that our ancestors endured. We may not have to wait long to find out.

While past shifts resulted from earth's climate adjusting to changes in incoming solar energy, and the dynamics of ice and oceans, it looks ever more likely that we may bring the next climate upheaval on ourselves. From the perspective of the solar-system dynamics that govern climate, it is difficult to imagine that puny humans could affect a climate system that was forged by the position of the planet relative to the sun, the movement of continents and the uplift of mountain ranges, the development of continent-sized ice sheets and planetary scale explosions. The other pieces of the heat distribution system that make earth a friendly place for humans, however, depend on tiny differences in the makeup of the atmosphere or subtle differences in the density of water. CO_2 makes up only .00000036 percent of the atmosphere. But if that changed to .0000036 percent, we would likely be living in the steam-room–like conditions of the age of the dinosaurs. The oceanic engine that releases 1,000 times the generating capacity of the United States in the North Atlantic is driven by subtle differences in the density of seawater moving through seawater.

Such are the intricate interconnections of this system that these seemingly insubstantial factors can offset the influence of giant planetary-scale climate influences such as the orbit and attitude of the earth. Belatedly, scientists have come to realize that humanity has been stomping through this minefield of invisible climate trip wires, and a great race is underway to understand what we have already done, and what lies ahead.

This brings us to the present.

PART FIVE

The Elephant in the Room

18

The Tides of Public Opinion

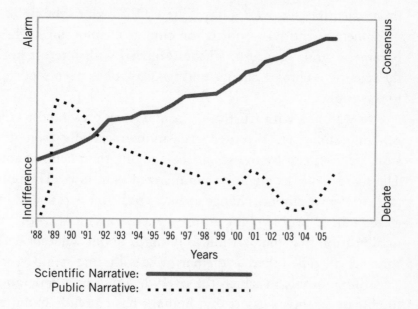

Scientific Narrative: ▬▬▬▬▬▬▬
Public Narrative: • • • • • • • • • • • •

WHEN PAST CIVILIZATIONS had to deal with climate change, they may have sometimes felt that they had brought calamity on themselves by offending the gods. Ancient scientists did not have the tools to understand long climate cycles. That is not the case today. Although scientists may lack complete knowledge of how climate changes, they can look back on the impacts of climate on past civilizations, understand some of the factors that drive climate change and what those changes might be, and monitor whether climate is changing. Since we live in a democracy marked by the free flow of information, we should be able to do something about it if we are causing the changes.

Any look at present-day climate change must take into account how information about the issue has been received by politicians and the public. So far in this book, I've looked at the debate about the role of climate in the fate of past civilizations purely in terms of the back-and-forth within the scientific community. Now let's step back and look afresh at how these scientific discoveries have filtered through to the public. As someone who has reported on climate change since it first became a public concern, I have followed with interest both the science of climate change and the way it has been reported to the public.

I've watched with frustration as the story presented to the general public has diverged ever more markedly from the story as it is seen by the scientists studying the phenomenon. The two narratives might be summarized as follows: The public story views climate change as likely to be moderate and incremental, a problem for future generations, and with the qualification that the best minds disagree about whether it's a threat at all. Then there is the narrative gathering steam in the scientific journals in which the overwhelming preponderance of climate scientists assert that humans have already had dramatic effects on climate, and that climate, when prodded, is prone to violent and extreme swings rather than gently paced changes. This paradigm shift (as it has been described by the National Academies of Science) has taken place with little public notice, but it is dramatic and the shift has occurred in little more than a decade—an extraordinarily short period for so profound a change in basic assumptions in a scientific field.

The implications of climate as an "angry beast" are deeply frightening, but both the public and the politicians remain oblivious, if not dismissive, of the threat. It's hard to imagine another issue with such important bearing on our future in which there is a larger disconnect between the scientific and public perceptions. In 1997, at a White House meeting on climate change, President Bill Clinton lamented that he could not take action on the issue unless the public took it seriously. Nearly eight years later the science is even more compelling but the public remains complacent. How did this happen?

As far as the public is concerned, the first serious warnings about climate change date back to 1979, when four distinguished scientists prepared a memorandum for President Jimmy Carter warning that human-caused emissions of greenhouse gases could cause "a warming that will probably be conspicuous within the next twenty years." From the vantage of 2005, the scientists were right on the money, but during the subsequent Reagan years, the administration had no interest whatsoever in climate change, nor any other environmental issue for that matter. When Senator Timothy Wirth of Colorado held hearings on climate change in 1987, the experts he'd invited spoke to a nearly empty room.

Then came the sweltering summer of 1988. More than 2,000 daily temperature records fell in the United States, and Washington wilted as temperatures topped 90 degrees for twenty-one straight days. This time when Wirth held hearings, there was a packed house to hear James Hansen of NASA sound his warning. When he asserted that humanity was already seeing a signal of global warming, the prospect of rapid climate change was regarded as little more than a radical hypothesis in the imagination of a couple of oceanographers who speculated that shifts in ocean currents might have massive and abrupt impacts on climate. Indeed, early climate models barely acknowledged the role of the oceans at all. Subsequent discoveries have revealed the oceans to be a sleeping giant in the climate system.

Unfortunately, most politicians and the public still live in

1988 in their understanding of climate change and their willingness to do something about it. That's because 1988 also witnessed the launch of the two distinct story lines on climate change, currents, as it were, that moved through the noosphere without mixing just like the warm and cold currents that can move through the Atlantic for centuries without mixing. Let's consider the public story line first.

The standard climate-change template for the national media usually begins with a peg—a collapsing ice shelf, a heat wave, retreating glaciers, devastating hurricanes—and then offers a scientist who ties the event to a warming globe. The story usually includes a recapitulation of the basic science (which eats up a good deal of the story), a bit on the many unknowns of future climate change, and then gives the naysayers a chance to dispute the notion that climate change is a threat. Most stories also mention the expense of taking action. I've read scores of these stories, and the takeaway message is that climate change is a complex problem with many unknowns and expensive solutions, a problem that won't impact our lives for many years if at all.

This public story line has evolved in response to a variety of factors, not the least of which is the widely shared attitude among editors and television news producers that the American public has a short attention span and little capacity to understand science. Another potent influence came from industry-financed lobbying groups such as the Global Climate Coalition and the Western Fuels Association. They sent forth their troops to spread the message that there is active debate within the scientific community about whether climate change is a threat at all. While scientists tended to exercise caution before making sweeping statements, the lobbying groups trumpeted the unsupported claims of their hired guns, and energetically issued press releases asserting that global warming was a myth or that if it was real that it would make the deserts bloom. Members of the media (including this writer since I wrote about global environmental issues for *Time*) found themselves hounded by experts who conflated scientific diffidence with scientific uncertainty, and who wrote outraged letters to the editor when a report didn't include their dissent.

It is small wonder then that major climate-change coverage in the national media has been timid and fitful. With so much space given over to the most rudimentary science and venting by naysayers, the public was left with the impression that there was active debate about the threat long after the scientific community reached consensus. Not that the scientists were very forceful in asserting that consensus.

Indeed, the real synergy in the climate-change story has been that of cautious scientists interacting with cautious policymakers, all to the delight of naysayers who hold that no action is necessary. For instance, a principal and entirely appropriate source for journalists has been the reports coming out of the Intergovernmental Panel on Climate Change (IPCC), a group convened following the so-called Earth Summit in 1992, which tried to forge a consensus of more than 1,500 scientists from more than sixty countries. The panel took what they considered to be the best available science and then hewed to the middle ground. Policymakers then negotiated on the basis of this compromise, a process that embodied the remarkable assumption that climate would play ball with the bureaucrats and meet them halfway. To paraphrase the eloquent Secretary of Defense Donald Rumsfeld, climate negotiators chose to fight an enemy they liked, rather than the one that is real. As is often the case with science, however, the new model of what climate change might bring and how fast came from the edge, not the middle, of the scenarios offered in 1995.

Given the forces buffeting the scientists, it is something of a miracle that the IPCC managed to make any forceful statements at all. In 1995, the group began circulating a draft document warning of dire consequences of a 6-degree Fahrenheit warming over the next century, which would be accompanied by a three-foot rise in the sea level, shifts in rainfall patterns, and increases in droughts and severe storms. A 1995 *Time* magazine article on the draft report (not written by me) presented these consequences as a worst-case scenario, but many, if not all the scientists in the IPCC knew even then that this was not anywhere close to a worst-case scenario.

The real worst-case scenario began to take shape in another narrative, and it has been gathering momentum in the scientific journals, publications such as *Science, Nature, EOS,* and the *Journal of Quaternary Science,* publications that for the most part lie beyond the reach of the industry lobbying groups that exert pressure on the IPCC. This narrative drew much of its initial energy from a $25 million National Science Foundation project launched in 1989 to extract a two-mile ice core from the center of the Greenland ice sheet. The goal of The Greenland Ice Sheet Project, or GISP2, was to obtain a record of climate going back tens of thousands of years.

It's a bit ironic that it took the study of ice cores on land to open the eyes of the climate community to the role of the oceans in climate. But with the GISP2 findings about abrupt changes, tumblers began falling into place in the most creative minds in climate science. Theoreticians and oceanographers started trying to explain both how such rapid change could have happened and what might have triggered it. This in turn brought oceanographers into the picture, since the capacity of the oceans to store, move, and release heat dwarfs that of the atmosphere. Wallace Broecker of Lamont-Doherty was one of the most imaginative oceanographers to take up the challenge. He had been waiting for evidence confirming his suspicions about rapid climate change since his days as a graduate student in the 1950s, and he saw the ice core results as an opportunity to test his ideas about the role of thermohaline circulation—the flux of salt and heat in a vast ocean current that stores and releases heat during the current's slow journey through the world's oceans—in rapid shifts in climate. As early as 1989, Broecker was publishing his suspicions that THC was the Achilles' heel of the global climate system.

In 1996, I met Broecker, Alley, Kendrick Taylor, Peter de-Menocal, and a number of other scientists in the thick of the question of rapid climate change. The National Science Foundation invited me to visit Antarctica as part of a fellowship program that brought journalists, writers, and artists to research sites. I seized on this as an opportunity to explore the increasingly wide gap between the vision of climate change that governed the IPCC and policymakers, and the very different picture of climate change that was gaining adherents in the climate community.

The IPCC report made glancing reference to the possibility of the weakening of THC and, in the section on uncertainties, had cautionary words about the possibility of rapid, "nonlinear" shifts and "unexpected behavior" in the climate system. The policymakers, however, fastened on the middle-of-the-road picture of a measured pace of change and moderate shifts. I was struck by the absurdity of designing a climate treaty predicated on the "dial" model of climate even as growing numbers of scientists had come to acknowledge that climate is governed by a switch. It was akin to looking for dropped keys under a streetlamp because that was where the light was best.

The subtitle of the article I wrote for *Time* read: "The conventional wisdom is that climate change will be gradual and moderate, but what if it is sudden and extreme?" Since I was in Antarctica, I focused on the Antarctic component of ocean circulation and its role in rapid climate change—it's awkward to dateline oneself near one pole and write about events at the other. Looking back, I can see that in some sense, I displayed the bias of geographical anchoring that afflicts many of the scientists working in the field. The problem is that the world is so large and climate so complex that some degree of synecdoche—mistaking the part for the whole—is all but unavoidable.

Within the next year, a number of detailed articles on rapid climate change appeared in the scientific press. Broecker published his article in *Science*: "Thermohaline Circulation, the Achilles Heel of Our Climate System: Will Man-Made CO_2 Upset the Current Balance?" In his conclusions, he wrote: "Through the record kept in Greenland ice, a disturbing characteristic of the Earth's climate system has been revealed, that is, its capability to undergo abrupt switches to very different states of operation. I say 'disturbing' because there is surely a possibility that the ongoing buildup of greenhouse gases might trigger yet another of these ocean reorganizations and thereby the associated large atmospheric changes. . . . More problematic perhaps than adapting to the new global climate produced by such a reorganization will be the flickers in climate that will likely punctuate the several-decade-long transition period." The article generated a furor in climate circles. The public and Congress, mesmerized by Bill Clinton's sex life, barely gave global warming a thought.

If knowledge was changing rapidly in some respects, its pace was still slow compared to alarming changes in climate. Indeed, the science might have changed even faster but for bureaucratic constraints. The four-year gap between the initial paper on rapid climate change and the next pulse of scientific papers seems slow, but Ken Taylor says that it reflects the pace of funding. He relates the history succinctly: "When Willi Dansgaard found a signal of rapid climate change in '83, the climate community said, 'Gee, that's interesting,' but then forgot about it because it was just one finding out there by itself. Then we found the same thing in GISP2 in '93, and scientists had to acknowledge that rapid change was real. Then a new cycle began with oceanographers and others taking a closer look at past changes to see whether there were other rapid shifts. That new cycle takes 4–5 years from funding to results." Taylor admits that the process is "maddeningly slow," and adds, "I'm afraid that we scientists will end up as historians—that people will look at our work and say, 'Oh that's why it happened,' rather than look at our work and say, 'Oh, so this is what might happen if we don't take action.'"

By the end of the millennium, the climate community was in the midst of a full-blown rapid paradigm shift. By 2000, many scientists in the field recognized that rapid and extreme shifts in past climate were the rule rather than the exception. Scientists were now finding evidence of past rapid shifts everywhere—in ice cores, in sediment cores taken from the sea bottom, and a host of other proxies.

In 2000, I traveled up to the Hudson Bay to write about how changing climate was already affecting the Arctic. The story I wrote for *Time* focused on the dramatic changes already happening as permafrost melted and sea ice thinned, and by going to the Arctic, I finally had a dateline that allowed me to more fully describe the role of thermohaline circulation in rapid climate change. Still, my article only scratched the surface, given the flood of new questions that accompanied the shift toward looking at climate as an angry beast.

Then came 2004, a pivot point in the story of climate change when many of the various threads in the scientific and policy debates converged. On the policy front, Russia ratified the Kyoto Treaty, guaranteeing that it would come into force. While by no stretch is this treaty adequate to the scale of the threat, it reduces the influence of the industry lobbying groups, since the United States can no longer stop international action. Instead of worrying about the costs of compliance with Kyoto, major corporations now face the threat of liability and trade sanctions.

The two narratives of climate change also began to converge, though the message coming from the media has been anything but consistent. On the one hand, the sci-fi film *The Day After Tomorrow* challenged the cozy assumption that we can take climate for granted, even though the film presented a preposterous scenario for rapid climate change. A Defense Department–commissioned study also drew attention because it specifically tried to envision the instability that might result from a rapid change in climate (Wallace Broecker, who had raised his own alarms about rapid climate change in *Science* in 1997, attacked the study in those same pages as alarmist).

On the other hand, the notion that climate change is a con cooked up by scientists looking for funding still seems to have as much traction with the media as the threat itself. In December 2004, delegates convened in Argentina to negotiate ongoing issues in the Kyoto Treaty. While this meeting got scant coverage in the national media, a new novel by Michael Crichton that dismisses the threat of global warming got massive publicity, including a fawning interview by John Stossel on the ABC newsmagazine *20/20*. Here's Stossel's lead-in to a discussion of the book: ". . . he's concluded [that global warming] is just another media-hyped foolish scare. And many scientists agree with him." Clearly, neither had read Naomi Oreskes's essay on the scientific consensus on climate change that appeared that very week in *Science*. The essay examined 935 peer-reviewed papers on climate change over a ten-year period from 1993 to 2003. Of the roughly 700 scientific papers that dealt with modern climate change, not one took issue with the consensus that humans are changing climate. It should be noted that a larger survey in 2004 of climate researchers revealed that about a quarter of the groups were still not completely convinced that humans were responsible for recent climate change, but this minority has not been expressing that opinion in peer-reviewed journals.

Ordinarily, such an overwhelming consensus in the scientific community would make me nervous, particularly since the consensus so often proves to be wrong. But it is not always wrong. The consensus on the link between cigarette smoking and cancer, for instance, seems pretty solid (Broecker gleefully told me that one of the few remaining serious skeptics on climate change, Richard Lindzen of MIT, also disputes the connection between cigarettes and cancer). Yes, there are many open issues in climate science, but with climate changing, ignoring the threat of climate change until every question is settled is a bit like refusing to run from an oncoming tsunami in Phuket, Thailand, simply because no tsunami had ever hit that coastline.

We find ourselves poised at a fateful point. We've diligently hauled ourselves to the top of a roller coaster with no idea how steep and long will be the plunge that awaits. It is true that no one can now say with certainty whether earth's recent sudden warming is a prelude to wild climate gyrations, or whether gyrations, such as a shutdown of the system of deep ocean currents that warms northern Europe and parts of North America, will be as extreme as past episodes in the climate record, but two vital components of the oceanic machine that delivers heat to the north have already partially shut down, and the climate system sends increasingly frequent distress signals.

19

Water Moving Through Water

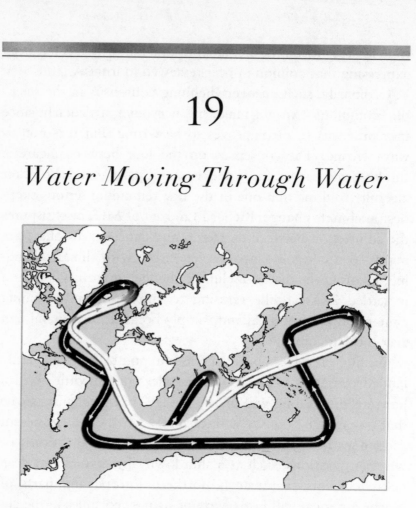

THE OCEAN CONVEYOR

TOWARD THE END of the summer of 2004, I had the opportunity to observe a team of oceanographers as they monitored the Gulf Stream and deep ocean currents, which have become the focus of many concerns about climate change. The cruise I joined was one of several in the Atlantic that summer. The struggle to understand the system is a little like an aquatic version of the *Maltese Falcon:* scientists chase around different parts of the North Atlantic, discarding false leads, looking for clues to the engines that drive this planet-scale system. The proliferation of efforts led one oceanographer to remark that "there are so many boats out there in the North Atlantic that I'm surprised they don't run in to each other."

To outsiders, the arcane worlds of climate modeling and theorizing might seem the sexiest part of the field, but some of the most important scientific work being done today is the most prosaic—the task of collecting measurements of the oceans and their currents. This need prompted Terry Joyce's latest research effort, now in its eleventh year. For this year's cruise, Joyce, an oceanographer based at the Woods Hole Oceanographic Institution, secured the *Cape Hatteras,* a 147-foot National Science Foundation research vessel.

I met up with Joyce at the boat at WHOI's dock in Woods Hole. Walking to the ship, I passed a weird-looking vessel sitting forlornly on the side of the pier. It was the *Alvin,* the institute's legendary deep-diving submersible. After six hundred dives, many more than 14,000 feet down, and forty years of service, the little sub is a relic of an earlier, more innocent era when scientists were more concerned with the exotic life forms of the abyss than whether the ocean might reach out and slam those far from its shores.

Our objective as stated on the Cruise Plan for the *Cape Hatteras* was to "Reoccupy Line W," which meant that we would travel about 600 miles from Woods Hole to within 100 miles of Bermuda and then stop at twenty-four stations marked by coordinates along the way in order to take measurements at different depths. Apart from Joyce, the scientific staff of the cruise included two scientists from Lamont-Doherty, technicians to analyze the retrieved water samples, and software and mechanical troubleshooters, including Brian Guest, a technical wizard who proved capable of diagnosing and repairing sophisticated equipment in minutes.

Our sensing devices included an array of high-tech bottles that can be tripped at various depths to collect water samples, and an underwater glider called *Spray* that, after launching, can fly slowly through the Gulf Stream for forty days, taking measurements and occasionally phoning home, before it is picked up off Bermuda. Similar gliders would be launched by a University of Washington team off Greenland to monitor another crucial part of North Atlantic circulation, while other teams of scientists, including consortiums from Canada, Germany, and the U.K., monitored other crucial parts of the circulation. These periodic measurements provided backup for moorings that WHOI and other institutions have anchored to the bottom, providing continuous measurements of current and salinity.

Terry had been to sea many times, and after dozens of cruises, he knew this ocean. Glancing at a graph of the water column, he reeled off the different currents merely by looking at changes in the oxygen content at various depths. "See, that's Labrador intermediate water," he said, pointing to a slight bulge in a graph of oxygen measurements below the surface. The purpose of these readings is to see whether anything is happening to the Gulf Stream and the other currents that help make Europe a nice place to live. "If the conveyor stopped," said Joyce, referring to the slang term for the THC, "we'd know about it in one cruise." The rest of us would find out a few years thereafter.

Joyce looks at the Atlantic as a series of currents and heat transfers, pointing out that as the Gulf Stream moves northward it constitutes a gigantic barrier, blocking the flow of "slope" water that flows off the continental shelf as well as the flows of intermediate water to a depth of 2,500 meters. He loves talking about the system, and he describes some of the ways in which the ocean "hands the ball off to the atmosphere." Water evaporated under the baking sun on the Sargasso Sea (it takes 540 calories to evaporate one gram of water, versus 1 calorie to raise a gram of water's temperature by 1 degree centigrade) forms clouds and then tracks the current northward, releasing the stored heat when the vapor condenses as rain over Europe. Thus two meters of evaporation of vapor off the surface of the ocean produces 1 meter of rain over the Sargasso Sea, but exports another meter of water to the Greenland Sea in the far north. There's an austere symmetry in Joyce's portrait of the ocean. It is utterly different from the perspective of, say, a Winslow Homer, but still beautiful.

The system is a testament to the subtlety and beauty of physics. It consists of something as diaphanous as water moving through water. The scale of the convection is tiny given the size of the oceans, and it would be invisible to someone right over it on the surface. It's too complicated for oceanographers to predict, and yet it is essential to the richest economies on the planet. That's the question: are we in for another Younger Dryas?

Fortunately, an exact repeat of the Younger Dryas is a very low-probability event. Unfortunately, it is not a zero-probability event. There are many unknowns. At the time of the Younger Dryas, great remnants of the ice sheets remained in the Arctic, providing a reservoir of fresh water. Where would that fresh water come from today? Some scientists, such as paleoclimatologist Daniel Schrag of Harvard, argue that the amplifier of the Younger Dryas was a great increase in sea ice, and that a shutdown without sea ice would cause only a modest and localized drop in temperatures. Others, like Joyce, are not so sanguine, and point out that even a so-called moderate drop would involve temperature during the most bitter winters of modern times, such as 1977–78. Temperatures that winter averaged about 6 degrees below normal, and Joyce took walks on Buzzard's Bay. He argues further that no one can eliminate the possibility that a shutdown would produce a great expansion of sea ice (keep in mind that in Antarctica during winter the sea ice expands at 30,000 square miles a day) and bring about a more catastrophic drop.

At present, no one knows whether we should be blasé or scared witless. No one really knows for certain how the ocean current system works, how it connects to the atmosphere, what might shut it down, or when it might shut down. As noted, scientists can't even agree what to name the system.

Then there is the question of when something might happen. There are varying timescales for the responses of air, water, sea ice, and glaciers to our alterations of the atmosphere. The atmosphere reacts quickly, but Tim Barnett of Scripps Institution of Oceanography points out that there might be a twenty-to-thirty-year time lag in the response of the oceans to the warming influence of greenhouse gases. The ice sheets are an even bigger mystery, conflating timescales that range from seasonal to millennial. In Antarctica's West Antarctic Ice Sheet, a pulse of warming that began at the end of the last ice age has been working its way toward the surface at the rate of about one foot a year. At the same time, some of the floating ice shelves that help keep Antarctica's ice in place have disintegrated in a matter of days.

In the peculiar world of ice dynamics, ice sheets can melt in one-tenth the time it takes to grow them, and as Richard Alley points out, big sheets are more vulnerable than small ones. For decades, scientists were blasé about the possibility of a catastrophic breakup of Greenland's ice sheet, which is about the size of the western United States, but recently this sangfroid has been challenged, as evidence of melting has started to show up in unexpected places. Oceanographer Wolf Berger of Scripps points out that since the Greenland Ice Sheet sits upslope, once the ice becomes vulnerable (from the heat introduced by meltwater) and starts to slide, a combination of gravitational energy and the heat from friction could lead to a runaway collapse. So it's not just ocean circulation that we have to worry about.

This is science in real time. The combustible brew of uncertainty and high stakes has distinguished scientists exchanging heated words, reversing long-held positions, and even contradicting themselves in the course of an interview. Underscoring this note of urgency is the discovery that two crucial parts of the system have already shut down. No one can say for certain whether such work stoppages are a normal occurrence or whether we are witnessing the first steps of a slowdown or shutdown that will plunge the world into climate chaos.

If the twin narratives of climate change have converged in recent years, so have the investigations of past and present changes in the ocean circulation. Paleoclimatologists like sediment specialist Lloyd Keigwin have begun looking at ever-more-recent events, while other oceanographers are trying to take the measure of present-day ocean circulation so that we will know when it begins to change—if that has not already begun.

As the story of thermohaline shutdowns has come into focus, new issues arise. Perhaps the most crucial for our future well-being is the question "Has the circulation already started to change?" It was this concern that prompted Terry Joyce's cruise, and those during the eleven preceding years.

We encountered the Gulf Stream on the second day, and our reception was not particularly friendly given that we had only the best intentions for its continued well-being. The current was moving northeast while the winds were coming from the northwest, setting up confused seas that bounced the *Cape Hatteras* every which way. The lumpy oceans led me to one of the little secrets of the world of oceanography: the surprising number of scientists who suffer from seasickness. I figured this out by deduction, since oceanographers are understandably loath to admit to seasickness. Many oceanographers really hate going to sea on research cruises. Not Joyce though, who has an iron stomach, and also seems to have missed the literature on skin cancer since he spends some of his free time basking in the sun, while occasionally getting launched into the air from his perch at the stern by the bucking of the vessel.

The guts of this mission was a 1,000-pound, $200,000 apparatus designed to withstand the pressures at depths of over 16,000 feet, where the deepwater flows. It contained the Doppler current profiler, another machine called a CDT that monitors conductivity (a proxy for salinity), density, and temperature, and twenty high-tech water bottles that can be tripped from the surface to collect water samples at various depths. Essentially, Joyce, and colleagues monitoring similar lines at different latitudes, are setting up border-control stations across the western boundary of the Atlantic Ocean through which no current can get through undetected.

Once the apparatus was safely on the deck, the technicians extracted samples from the bottles. The crew drew water to analyze for oxygen and salt concentrations, while a team of scientists from Lamont-Doherty under the direction of oceanographer William Smethie ran their samples through an onboard gas chromatograph to look at concentrations of chlorofluorocarbons and iodine-129. Because both chemicals are modern and man-made (the iodine comes from nuclear waste dumped into the English Channel by the French and British), they serve as useful tracers as they move through the oceans.

The work was constant and repetitive. Dave Wellwood, a research assistant at Woods Hole, walked me through the process by which he titrates water samples to determine oxygen levels (the scientists monitor oxygen levels to determine how recently the water has been at the surface). He placed a water sample in a machine where iodide is added. Then the machine adds sodium thiosulfate to liberate the iodide, and the amount needed to get rid of the iodide gives a very close (1.1 to 1) approximation of the amount of oxygen in the sample. After running a few samples through the machine, he looked up with a modest smile and in the self-effacing manner of techies everywhere said, "Basically a monkey could do this."

There was a lot of downtime, particularly when paying out three miles of cable as we dropped the apparatus to the bottom of the abyssal plain. A leitmotif of conversation was the relative merits of various research vessels. The *Cape Hatteras* gets high marks for its crew and food (Bob Lipscomb, the chef, one night served fresh sushi from a yellowtail tuna that Brian caught off the stern).

Joyce didn't have to be on the boat—most often chief scientists send reliable postdocs out to monitor the actual research—but he seemed to enjoy it. During a break, he told me how abrupt climate change and the role of the oceans came to be the central mission of his career. He had first come to WHOI as a graduate student in the late 1960s, an original member of a joint MIT-WHOI program in physical oceanography. At MIT, he studied with oceanographer Henry Stommel and at WHOI he'd had many encounters with the late physical oceanographer Val Worthington, so he had early exposure to ideas about the role of salty, dense water in ocean circulation.

Scientific disciplines tend to be pretty insular, and when Joyce first began to wonder whether the oceans might play a role in abrupt climate change, he knew nothing of the work being done along the same lines by paleoclimatologists. In the early 1990s, he began a study sponsored by a program called WOCE (World Ocean Circulation Experiment) to review data on Atlantic Ocean currents and establish some benchmarks against which change might be measured. Some of that data go back as far as the International Geophysical Year in 1957, while other data went back to the 1920s. Where he had comparable data, he noticed a trend: over the years the deep ocean currents had become both warmer and less dense. He interpreted this data as an indicator of global warming, "a subsurface indicator that we were warming deep water."

Around that same time, Joyce did a slightly more sophisticated version of a model developed by the late Henry Stommel that showed that changing density and salt can bring about a catastrophic shutdown of thermohaline circulation. He published his results in the *Journal of Geophysical Research*. That got him thinking about abrupt climate change. "At that time I didn't know anything about Richard Alley's work, but I started talking with Lloyd Keigwin, who was also at WHOI." Keigwin had done pioneering studies of ocean sediments in the Bermuda Rise that provided evidence of abrupt climate change during the glacial era.

Consequently, over the years his work shifted slightly. While still developing data on mean ocean currents, Joyce and colleagues also began to focus their efforts on a monitoring system for the Deep Western Boundary Current, which flows down the east coast of North America off the continental shelf and is the principal way the densest water formed in the Arctic leaves the region. It's an interesting project, because U.S., German, Canadian, and British scientists have come together to do this without ever having an international meeting to plan the effort. All are parties interested in the vigor of thermohaline circulation, and as Joyce puts it, "changes in the Deep Western Boundary Current would provide an early warning system of changes in the THC."

Ruth Curry, also at WHOI, has been looking at another critical part of the picture—an underwater ridge between Greenland and Iceland (incongruously named the Denmark Strait), over which the deepest and densest water flows before continuing its journey as the Deep Western Boundary Current.

Once deep water forms by sinking in the far northern seas, the densest and coldest water meanders toward the west coast of Greenland following the contours of the bottom, dropping as low as 12,000 feet before being forced to rise as the bottom rises toward the Denmark Strait. As more dense water feeds in, the water piles up and then flows over the strait's ridge, which lies about 2,000 feet below the surface. Once it has cleared the ridge, the water then sinks again, pulling other dense deep water behind. The amount exiting the Arctic in this flow going south roughly equals the warm flow entering the Arctic at the surface.

The so-called "Denmark Strait Overflow" offers a key monitoring point in the system because the engine that draws off the mountain of deep water piled up behind the ridge is the difference in density between water north of the ridge and that south of the ridge. In essence, the ridge is like a hydraulic control of the conveyor. So long as the water on the north side of the strait remains denser and colder than the waters to the south of the ridge, the flow will continue. Ruth Curry, however, has noticed a steady freshening and lightening of the water. In other words, the difference between the waters on either side of the ridge has been steadily diminishing.

Given the rate at which the density differences have been diminishing, Curry believes that it will be another four or five decades before the densities equalize on either side of the ridge and the overflow peters out. She acknowledges that that prediction is an educated guess that could be changed by a number of factors, not the least of which is that changes seem to occur because of tipping points and thresholds. No one now knows what these may be. Wally Broecker guesses that it would take a good bit of further warming—he estimates about 4 degrees centigrade—before the THC might shut down, but admits that he could be surprised.

The Arctic regions have warmed far faster than the rest of

the world, and as precipitation increases, runoff into the polar regions has increased. The increased rainfall has poured a lid of fresh water into parts of these oceans. Can precipitation and melting put enough water into the North Atlantic to shut down the THC? If it shuts down, will the cooling produce sea ice that amplifies the cooling, or will the making of sea ice sequester fresh water and restart the circulation (as some scientists claim)? Someday we will find out, since right now, no one knows.

In the meantime, other portents are not encouraging. Ruth Curry notes that the deep convection in the Greenland Sea remains shut down. There is still convection there, but the less dense waters are only sinking to intermediate depths. Similarly, deep convection in the Labrador Sea has not been observed since 1998. There too, water is still sinking to intermediate depths. No one knows whether such shutdowns occur regularly or whether we are seeing the beginnings of a shutdown of the whole system. I'd feel a lot better if this deep convection was still chugging along.

The phrase "we don't know" pops up continually in climate, and as Ken Taylor points out, the pace at which these holes in our knowledge become filled is as glacial as the ice he studies. Those who balk at taking action to address the human contribution to climate change assume the cloak of prudence given scientific uncertainty, but it should be clear that climate is so complex that certainty will only come in retrospect. The world knew enough about the potential dangers of climate change to take action sixteen years ago, and even the milquetoast actions required by Kyoto won't bite for another few years.

If the pace of scientific work remains slow, the number of studies has exploded. As scientific energy has focused on the role of THC in the North Atlantic climate, some tensions have surfaced about the best way to portray the risks.

During free moments, I talked with Terry about the controversy stirred up by some of the public warnings about the threat of a shutdown of THC coming from Joyce and other scientists at WHOI. More through coincidence than design, WHOI has achieved a high profile on concerns about the consequences of THC shutdown. Joyce wrote an op-ed piece for the *New York Times* in 2002, warning about the potential consequences of a shutdown. That same year, WHOI director Robert Gagosian delivered a warning on abrupt climate change to the World Economic Forum in Davos, Switzerland, and he also cooperated in the Pentagon study released in 2004 that offered a worst-case scenario of the consequences of a shutdown.

Earlier, Daniel Schrag had told me that he considers some statements coming out of WHOI irresponsible. He asserts that at best THC delivers only 15 percent of the heat to northern Europe and that without the amplifier of sea ice a shutdown would not be a major catastrophe. When I mentioned Schrag's demurral to Joyce, he responded that the amount of energy depends on the latitude, with the mid-latitudes getting as much as 25 percent of their energy from the Gulf Stream, which in the past has shifted south during shutdowns. Moreover, he says that even without sea ice, a shutdown could lower temperatures 10 degrees Fahrenheit, making frigid winters like that of 1978 par for the course for hundreds of years.

Joyce also asserts that statements coming from WHOI are not always reported responsibly He notes that critics castigated him for writing in an op-ed that a new chill would start within the next ten years when what he wrote in fact was that changes would be felt within ten years of a shutdown. Given that the stakes could be civilization itself (there, I'm doing it myself; providing evidence of how easy it is to be an alarmist), I sympathize with his struggle to strike the proper note.

Having finished taking samples at various checkpoints, we

headed home toward Woods Hole. As we left the impossibly blue waters of the Gulf Stream, the surface temperature monitor plunged from 28 degrees centigrade (82.4 Fahrenheit) to 24 centigrade (71 Fahrenheit) in a matter of yards. That four-degree difference represents an enormous amount of energy, a portion of which gets delivered to the north courtesy of the peculiarities of the Atlantic Basin.

On the last two days of our cruise, Brian started final preparations for deploying *Spray,* the underwater glider. Run by oceanographer Breck Owens back at Woods Hole, the project will monitor the amount of heat moved by the Gulf Stream. Spray's job is to profile the upper ocean, porpoising up and down between the surface and a depth of 3,000 feet between Woods Hole and Bermuda, and occasionally phoning home with its readings

The ingenious machine was first envisioned more than twenty years ago by the late MIT oceanographer Henry Stommel. The drone is something like an underwater, battery-powered hang glider, with its wings providing lift and its direction controlled by shifting its ballast forward and back and side to side. It rises or falls by pumping oil from inside the glider to external bladders and vice versa. As the bladders inflate outside the fuselage, they change the displacement of the glider, which causes it to become more buoyant. Shift the oil back inside of *Spray* and the glider sinks. With no propeller, the little 6-foot-long glider is pretty close to a perpetual motion machine.

Spray looks a bit like a bright orange torpedo (and indeed I wonder whether we are being monitored by Homeland Security since we launch the glider on September 11), but it is a minutely calibrated piece of equipment. The launch required particular circumstances—daylight, calm seas, and water dense enough that *Spray* can get back to the surface after its initial dive. Unfortunately, by the time dawn broke on our final day, the water seemed too "light" for a launch. Terry called Breck, who in turn called the designer, Jeff Sherman, at Scripps (at 3:30 A.M. his time) to gin up a fix. Temperature wasn't the problem since the glider will expand to compensate, but we needed to find a place with relatively high salinity. After a quick calculation, Terry directed the captain to take us back to an earlier station on the line we have been following.

As we send the glider on its way, I'm in awe of the subtlety of this benign missile that was launched on September 11 to help us understand a threat far more profound than anything Osama bin Laden might conceive. *Spray*'s situation is a little like that of the thermohaline circulation itself: With water of the right density, the whole system works. Change the density, and it gets lost and confused. Once the glider has been lowered into the water, it floats for a few minutes as though it doesn't know what to do. Then it rolls and raises one wing to get its bearings, rolls the other way to beam its position to Woods Hole, and then, as though waving goodbye, rolls back to level before disappearing beneath the waves on what I very much hoped would be an uneventful journey through the Atlantic's principal artery.

PART SIX

Closing Arguments:
Are We Next?

20

Going Forward

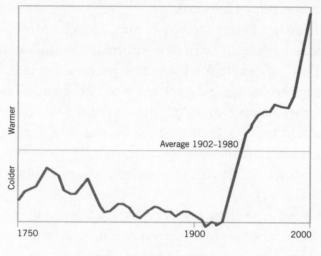

Warmer

Colder

Average 1902–1980

1750   1900   2000

TEMPERATURE CHANGE SINCE 1750

WHERE ARE WE HEADED? Is climate changing? If so, are we causing these changes? What changes lie in the future? Are we better prepared to deal with climate change? Can we do anything to halt climate change or ameliorate its effects?

We humans are particularly poor at envisioning weather different than what we are presently experiencing. If it's summer, it's hard to imagine a snowy day in winter. It's harder still to imagine climate turned upside down. It has been our good fortune to prosper and multiply during one of the most benign climate periods in the record. The great expansion of the Industrial Revolution, the sixfold increase in human numbers, the triumph of the consumer society and integrated global economy took place in really good weather. There have been destructive El Niños, but the last time the weather was as good as it was between 1825 and 1980, the wheel was a novelty in much of the world. And, of course, one would have to go back 115,000 years to find a period as tranquil and warm as the present-day Holocene.

We have not been tested by climate change. Moreover, humans have a tendency to fit new information into familiar patterns. This may explain why so few people have noticed that climate began changing during the past two decades, and even fewer have become alarmed. Just as tourists on a Phuket beach could make a fatal mistake by standing and gazing at an oncoming tsunami because it was outside their experience, so does society react to the coming wave of climate change without urgency.

The weather *is* changing. The last decades of the twentieth century saw an unmistakable and extraordinary warming. Almost all the warmest years in the modern temperature record have taken place since 1980, and during the nineties, the pace at which the warming has eclipsed earlier records has accelerated. In the summer of 1988, 2,000 temperature records were broken in the United States. In the early 1990s, the eruption of Mount Pinatubo in the Philippines put enough ash into the atmosphere to cool the air for a couple of years, but then global temperatures soared again and the records began falling rapidly. Nineteen ninety-five broke the record for warmth. Then 1997 became the second-warmest. The next year, 1998, broke the 1995 record. In the U.S., 1999 came in as the second-warmest year on record, and the fifth globally until it was supplanted in that spot by the year 2000. The very next year then replaced 1997 as the second-warmest year on record, until it was knocked down a notch in 2002, and knocked from the third spot by 2003. Then 2004 came along and knocked 2001 out of the fourth spot. And so it continues.

During this same period, we suffered a swarm of "droughts of the century," "500-year floods," what some described as the strongest El Niño in 130,000 years, and other weather extremes. On December 7, 2004, Tropical Storm Odette hit the Dominican Republic, the first recorded tropical storm to hit the Caribbean in December. That same day a monster snowstorm paralyzed a 750-mile swath of the U.S. Northeast, pummeling Maine with 82-mph winds and dumping 20 inches of snow on New York City. Two storms with hurricane-force winds—one cold and one warm—pummeled contiguous regions on the same day. Two days later, Tropical Storm Peter formed in the eastern Atlantic, the first time two such storms had formed in December. The year went into the record books as having witnessed the earliest and latest tropical storm activity on record. Not long thereafter, a hurricane hit Brazil, the first recorded hurricane ever in the South Atlantic.

We are familiar with the importance of some of nature's thresholds. A 1-degree variance in temperature can mean billions of dollars in damage if the variance is from 32 degrees Fahrenheit, and it is accompanied by precipitation along the east coast of the United States. We are about to become familiar with some other thresholds. Terry Joyce points out that when ocean waters rise above 82.4 degrees Fahrenheit, evaporation increases dramatically, providing an enormous reservoir of potential energy for storms.

This does not mean necessarily that hurricanes will become more frequent, but research led by Kerry Emmanuel of MIT suggests that hurricanes have doubled in intensity during the past thirty years as the oceans have gradually warmed. Hurricane Katrina surged to its immense power when the storm passed over a deep layer of 90 degrees Fahrenheit water in the Gulf of Mexico.

Nature does not alert us to all her trip wires. Perhaps that's why in recent years the unprecedented has become increasingly ordinary. Tornadoes hitting the Deep South in January 2005 had meteorologists scratching their heads. There was a precedent for the massive snowstorms that dumped up to 19 feet of snow on the Sierra Nevada in just a few days between Christmas 2004 and the beginning of 2005, but it was in 1916.

Are humans the cause of these events? The same week the storms afflicted the Atlantic region in 2003, *Science* published "Modern Global Climate Change," by Thomas Karl of the National Climatic Data Center and Kevin Trenberth of the National Center for Atmospheric Research. The scientists argued that "modern climate change is dominated by human influences," and that "we are now venturing into the unknown with climate, and its associated impacts could be quite disruptive." A year later, *Science* published Naomi Oreskes's essay "The Scientific Consensus on Climate Change," in which she listed the who's who of distinguished scientific groups that have concluded that humans have contributed to the observed changes. These groups include the IPCC, the American Meteorological Society, the American Geophysical Union, and the American Association for the Advancement of Science (AAAS).

You don't have to be a geochemist to get a sense of how humans are affecting climate. Look at three charts that have been published in almost every major newspaper and many national magazines during the past decade. One chart shows the rise and fall of CO_2, another the rise and fall of methane (a potent greenhouse gas), and another, global temperature variations from the present. It shows that when methane and CO_2 are up, temperatures rise, and when they fall, temperatures fall. The astonishing thing about the chart is that present-day levels of methane and CO_2 in the atmosphere are literally off the charts, which go back 400,000 years. Methane levels are nearly triple the levels during the Eemian, the last warm period, which ended 115,000 years before the present. CO_2 is up about 50 percent, but, as we saw, headed for a doubling or tripling. What's more, the rate of increase in CO_2 has sped up in recent years. Paul Brown of the *Guardian* quotes Charles Keeling, the climate scientist who first began taking readings of CO_2 on Hawaii's Mauna Loa in 1958, as saying that it's possible that this rise could represent "a weakening of the Earth's carbon sinks, meaning that the earth is increasingly losing it's ability to store the carbon we release into the atmosphere." Because temperature variance is judged against the present, this is the one chart that shows no change. But if that set of charts tells us anything, it tells us that temperature will change.

Are we better prepared to deal with climate change than our forebears? The short answer is that we may be, but nature isn't, and that means that we won't be either. During the heat wave in Europe in the summer of 2003, more than 35,000 people died, an astonishing figure given that the deaths took place in some of the richest nations on the planet (a British study estimated that by 2040 half of Europe's summers could be as hot as the summer of 2003). Until August 2005, Americans could observe such death tolls with some detachment. Then Hurricane Katrina brought home the destructive potential of extreme weather events. On the other hand, as we have noted, no El Niño in recent history has produced victims in numbers comparable to the tens of millions who perished during the Victorian-era events described by Mike Davis.

Still, no modern industrial society has been tested by the protracted climate chaos that destroyed the Akkadians, the Mayans, or even the Norse in Greenland. Nor has the global system of markets and food distribution, which has largely eliminated death by starvation except in Africa. The global grain surplus is vanishing, however, and it is open to question how the markets would respond should crops fail in the handful of food-exporting nations while synchronous famines gripped Asia and Africa.

There are no models to estimate the economic impact of rapid changes in temperatures, storm tracks, precipitation, etc. As the National Academies of Science report "Abrupt Climate Change: Inevitable Surprises" points out, most of the modeling of impact has been confined to cases where climate changes are gradual and moderate.* Most of those studying impacts are working from a picture of climate change that may have little relevance to how climate changes in the future. Modeling the results of abrupt climate change is a lot harder, but the National Academies study makes a couple of important points.

Studies of gradual climate change suggest that economies can minimize the shocks if societies recognize the threat and adapt. With abrupt climate change, however, things begin to happen so rapidly that neither societies nor ecosystems have time to adapt. Moreover, a dynamic market economy that has the capacity to spread risk and respond to intermittent crises may not have the capacity to respond when crisis is ubiquitous and risks increase everywhere.

In such cases, a situation can develop in which all the players attempt to reduce their risk and the markets begin to freeze. The world had a preview of the abruptness in which such paralysis can occur with the seizure in the financial markets when Long-Term Capital Management collapsed in 1998. One hedge fund's cash problems briefly froze global financial markets as institutions tried to reduce their exposure. The risks of abrupt climate change would not surface at the

*Committee on Abrupt Climate Change, National Research Council. *Abrupt Climate Change: Inevitable Surprises* (National Academies Press, 2002).

warp speed of financial markets, of course, but just as LTCM represented an unpredictable rogue wave that rose up to threaten the financial world, the waves propagated through the markets by myriad responses to the effects of changing climate would likely cause market failures throughout the global economy.

Here is a relatively simple example of how even moderate climate change might affect a particular area. Tim Barnett, the Scripps oceanographer, participated in a study of the likely effects of climate change on the Los Angeles area. They discovered that even modest decreases in rainfall during what he called a "best-case scenario for future climate change" could reduce available water for the area by 50 percent by 2050. It was a best-case scenario because it assumed relatively slight and gradual climate change.

As Barnett explained, the region has limited storage capacity for water and relies on the snowpack that builds up in the winter both for storage and to meter out the water during the dry summer months. Under even modest climate-change scenarios, however, the winter snowpack would be smaller and it would melt earlier. The region would dry up before its driest months.

Since 88 percent of the water in California goes to agriculture, supplies could be diverted for immediate human needs. The ancillary effects would be harder to manage. Drought stress makes trees more vulnerable to pests such as the pandora moth, which afflicts ponderosa pine, or the various beetles that eat other softwoods. Dead trees become tinder for wildfires, such as the ones that destroyed hundreds of homes in Southern California in 2003. Diseases liberated by the changing weather could also afflict crops, livestock, and even humans, as was the case when El Niños led to outbreaks of hantavirus in the Southwest in 1990s.

Indirect effects of drought surface in many ways. By 2005,

long-term drought in the catchment area of the Missouri River was threatening hydroelectric production. This was a predictable result since hydro requires water flow to drive the turbines. The drought also threatened electrical power generated by eighteen coal-fired plants downstream. They don't need water to drive the turbines but they do need water to cool them. A 2005 NPR report on the drought made the point that even though power generators could theoretically buy power from the national grid, as a practical matter such purchases are not always sufficient to meet demand, and that if those plants were to go offline, the regions served would suffer higher prices, brownouts, or both.

In the case of California, the richest state in the richest nation on earth, the government could probably respond to its challenges at some economic cost. Such changes, however, would not be occurring in isolation. How would increased fire risk and other derivative effects of acute water scarcity affect the job market or the real estate market? In Southern California, many homeowners have very little equity at risk because banks have been willing to finance nearly all the costs of buying homes. Will banks continue to do so if prices start to stall, or if insurance companies balk at insuring homes in high-risk areas? And what would happen to the banking system if banks become suddenly saddled with a huge increase in non-performing mortgages and unsalable properties possessed through foreclosure? With no cushion and no buyers, foreclosures would quickly propagate back up through the financial system. Because mortgages have been sliced and diced into so many derivatives, the crisis could quickly become systemic as investors fled markets.

Keep in mind that the 50 percent reduction of available water that Barnett and his colleagues modeled was a *best-case* scenario. The changing climate that brought drought to Southern California would also be affecting weather throughout the American West, the rest of the country, as well as every nation on earth. The events that produced a 50 percent reduction in available water for Los Angeles would likely produce even greater drought in Mexico, according to climate models. This would aggravate desertification, which in the 1990s drove about 1 million people off the land annually. Some portion of the dispossessed of Mexico and other regions of Central America would become illegal immigrants in the United States, further burdening an overstressed economy and infrastructure.

Changing climate will test nature as well as markets and economies, and impacts on nature will reverberate back to markets. While the development of markets and technology has better equipped us to deal with external shocks, the impact of our wealth and increase in human numbers has commensurately reduced the ability of natural systems to adjust to climate change. In turn, nature's decreased resilience directly increases the costs and impacts of climate change on humans. The consequences of Mike Davis's "ecological impoverishment," the diminished capacity of the natural systems that support economies, may someday apply to the rich as well as the poor.

Consider the "500-year" floods that caused $26.7 billion damage (in 2002 dollars) in the midwestern United States in 1993. Decades of projects had channeled and otherwise altered the great rivers of the Midwest, reducing the rivers' access to floodplains, which had absorbed and moderated the effects of extreme rainfall. The rivers rose higher than they might have in years past. When they breached dikes and other barriers, they spilled into floodplains now largely occupied with farms and homes, increasing the impact. (A total of 50,000 homes were damaged in the '93 floods.) In other parts of the world, such as China, Central America, and Africa, deforestation, rather than the loss of flood plains, has been the agent magnifying the effects of floods.

These events underscore the importance of tipping points. When pushed past a certain threshold, the damage of natural events increases exponentially. If natural buffers have been eliminated, that threshold is lower. The difference between a Red River crest of 49 feet (as was predicted for Grand Forks, North Dakota, during a 1997 flood) and 54 feet is 10 percent, but the economic damage resulting from a 54-foot crest might be more than a hundred times greater because the flood would breach containing barriers built under the assumption that the river could never reach that level. This was a reasonable assumption, because the river had never before crested that high—the peak volume of the 1997 flood at Grand Forks was twenty-seven times average flow and more than 50 percent higher than the "100-year" flood of 1979. Thus, efforts to contain natural events such as floods, combined with our conversion of natural buffers, has the triple effect of diminishing nature's capacity to modulate floods, increasing river levels, and dramatically increasing potential damage.

It was such modifications that exaggerated Katrina's impact on the lower Mississippi. Measures to streamline river traffic and protect New Orleans from floods contained the river, but deprived the flood plain of replenishing silt. As New Orleans and the surrounding bayous settled, the city became ever more vulnerable to flooding should the protective system of levees fail. At the same time, failure became more probable because the disappearance beneath the waves of thousands of square miles of wetland removed a protective buffer that might have moderated winds, making it more likely that the levee system would be tested by a storm's most punishing winds and surges.

Katrina also provides the starkest example of the exponential rise in human and economic costs when an incremental increase in intensity breaches defenses that have been designed for only slightly less damaging storms. To be sure, human shortsightedness before the storm and ineptitude afterwards exacerbated the carnage, as did chance events such as a barge breaking loose and breaching a crucial levee. But the fact remains that a storm only marginally stronger than previous hurricanes to hit the region resulted in death tolls vastly greater than recent strong hurricanes had caused and damage estimates measured in the hundreds of billions rather than the hundreds of millions.

Around the world humanity has reduced nature's capacity to dampen extremes in countless ways. The destruction of half the world's mangroves has removed a natural buffer that diffuses the effects of waves and reduces coastal erosion during hurricanes. From Afghanistan to Africa, deforestation has dried out regions and provided tinder, exaggerating the effects of wildfires. The statistics that document nature's diminished capacities are astonishing given the short period during which the changes have occurred: more than 59 percent of the world's accessible land degraded, half the world's available fresh water now co-opted for human use, half the world's wetlands drained or destroyed, one-fifth of the world's coral reefs destroyed and one-half damaged, and on and on. The negative consequences for ecosystems have been well documented, but they have eclipsed an important implication. Even as human activities have increased the likelihood of extreme weather events, they have decreased nature's ability to contain the damage. This has shifted the burden to us, and we're balking at paying to avert or adjust.

The visiting Martian observing the situation would conclude that humans have a self-destructive streak. Having proven through history that extreme climate shifts are costly if not fatal for past civilizations, we are busily pushing the climate toward producing extremes never before experienced during modern times. Instead of trying to avert climate change, we are dismantling natural protections against climate extremes and making changes that will likely amplify the events themselves.

The explanation, of course, is not masochism, but a combination of a collective failure of the imagination, and a belief in the almighty power of markets and progress. The threat of climate change lies so far outside our experience that we don't know how to weigh it, and that saps our political will to do something about it. Moreover, so many threats have either failed to materialize or have been averted by technological change that many Americans simply accept that this will happen in some mysterious way. The devoutly religious who believe that their fate is totally in the hands of God can't hold a candle to the faith of free marketers in their worship of technological progress.

Although alarms about the threat of climate change have done almost nothing to slow emissions of greenhouse gases or to prompt efforts at adaptation, there have been some gestures in the direction of change. Since climate began to go haywire in the 1980s, some most likely to suffer from the effects have begun to calculate how to spread risk and pass on costs. The insurance industry, for instance, could be crippled by claims if underwriters don't figure out how to price the risk of climate change and pass it on to customers.

A look at that industry's efforts to come to grips with climate change opens another window on the way costs will spread through the economy. Insurers have been looking at climate change since the early 1990s, when a series of windstorms and hurricanes burdened the reinsurance industry (reinsurers insure insurance companies against catastrophic losses) with claims that forced several firms out of business and almost brought down the venerable Lloyd's of London. It's difficult for insurers to manage risks such as climate change because, typically, a company insures a particular property for a particular period, and depending on where that property is located, the marginal additional risk caused by climate change is exceedingly difficult to measure.

A good analogy is the risk of terrorism. Before the 9/11 attacks, insurers were aware that terrorism posed a risk but in most cases assumed the risk for free because there was no efficient way to measure how to apportion that risk to a particular building, especially in the United States, which had largely escaped terrorist attacks. Swiss Re, one of the insurers of the World Trade Center, lost roughly 3 billion Swiss francs in those attacks. The company's analysts had investigated scenarios in which a plane might accidentally run into the buildings, but they had not anticipated that someone might intentionally crash a fully fueled plane into the towers at extremely high speeds. Their rates did not reflect the potential damage of the type of event that actually occurred.

Following 9/11, insurers stopped writing policies that included coverage of terrorist attacks, and a number of big construction projects temporarily halted because banks would not assume the financial risk. Ultimately, the Congress passed and President Bush signed a law that shifted responsibility for $100 billion in damage coming from future terrorist attacks to the U.S. government.

The 9/11 attacks opened insurer's eyes to other risks they had assumed for free, climate change being among the most obvious. John Dutton, dean emeritus of Penn State's College of Earth and Mineral Sciences, has estimated that $2.7 trillion of the $10 trillion U.S. economy is susceptible to weather-related loss of revenue, meaning that an enormous number of companies have "off balance sheet" risks related to climate. If climate change starts inflicting losses, insurers will again head for the exits. Just such insurer flight has already caused problems in North Carolina's Outer Banks and in parts of New

York's fabled Hamptons, where coastal storms threaten. When insurance companies quit these high-risk places, the burden shifts to those who make the loans (since homeowners and businesses typically have less than 50 percent of the equity of a property). Banks don't have as much freedom as insurers to cancel mortgages and loans, but they can put conditions on new lending. What will happen to the markets if banks start demanding insurance for weather-related events that is either prohibitively expensive or unavailable?

The climate-change threat that will really get the attention of executives and company directors is the possibility that they might be liable for damages. This could happen if insurers like financial giant Swiss Re start changing the insurance policies that insulate directors and officers (called D&O insurance) from the costs of lawsuits resulting from the actions of their corporations. Businesses open themselves to lawsuits when they take a position contrary to others in their industry, and in recent cases such as asbestos litigation, courts have assessed damages proportionate to a company's contribution to a problem.

Chris Walker of Swiss Re describes how this might come about with regard to climate change. He notes that energy giant ExxonMobil accounts for roughly 1 percent of global emissions and has aggressively lobbied against any efforts to reduce greenhouse gases. "So," said Walker, "we might then go to them and say, 'Since you don't think climate change is a problem, we're sure you won't mind if we exclude climate-related lawsuits and penalties from your D&O insurance.'" In 2004, Swiss Re set the stage for such action by sending a questionnaire to its D&O customers inquiring about their company's strategy to deal with climate-change regulations.

For insurers, the risks of climate change become more concrete each year. Andrew Dlugolecki, a leading risk analyst now at the Tyndall Center for Climate Change Research in the

U.K., recently estimated that, if climate gradually warms, the chances of catastrophic weather-related losses rise from about one in a hundred to nine in a hundred over the next fifty years. A ninefold increased risk of catastrophe obviously places a large burden on the insurance industry, but the risks may be far greater than that. When asked in 2003 how abrupt climate change might change those odds, Dlugolecki estimated that in a scenario of abrupt and extreme change, the risk of catastrophic weather-related losses rises to about nine chances in one hundred by as early as 2010. To insure a property or business most likely to be affected by this type of risk, a carrier would have to charge rates as high as 12 percent of value annually. To put that figure in perspective, most people and businesses start self-insuring when premiums reach 3 percent of value.

Why hasn't this happened? Broad risks get shoved to the margins when there is no available body of past data. If climate continues to change, insurers will pull back and otherwise offload increased risk. As the financial system and businesses pass these added risks to governments and individuals, it is likely that climate change will, at last, become a political issue.

By then, it will be too late. Because CO_2 lingers in the atmosphere for decades, it behooves governments to control activities that might cause climate change before it worsens. Now, with climate already warming dramatically, we are likely stuck with some form of climate change.

Wallace Broecker argues that we are still a ways off from sufficiently warming the earth to shut down THC, but at the same time he admits that no one can say for certain what the tipping point might be for a change of state in the climate system. It should be obvious that we don't want to push climate over the edge, but as of 2005, the global economy is still in the process of squeezing rather than releasing our pressure on the trigger.

With ratification by the Russians in the fall of 2004, the Kyoto Treaty achieved coverage of the 55 percent of 1990 emissions that was set as a minimum threshold for the treaty to go into force. The treaty mandates that industrialized nations reduce their emissions 5 percent below 1990 levels between 2008 and 2012. Between 1990 and 2000, industrial nation emissions had dropped by 3 percent, a statistic that by itself would suggest that the developed nations were aware of the threat and taking action. A closer look, however, reveals that the lion's share of that reduction was the result of the closing of antiquated and inefficient coal-fired industries in Russia in the years following the collapse of the Soviet Union. With Russian data excluded, industrial emissions rose 8 percent in that period. In fact, only a few EU nations will meet their targets, and industrial nations will likely be 10 percent above 1990 levels by 2010. Worse, from the point of view of meaningful action, is the fact that the United States has refused to sign Kyoto, and the world's second-largest emitter of CO_2, China, is not subject to its provisions.

The Bush administration pays lip service to the threat of climate change and has offered a plan to reduce carbon intensity (the amount of CO_2 relative to a unit of production) rather than overall emissions. By any measure, the program is a mockery of serious commitment. For one thing, carbon intensity has been dropping anyway as businesses adopt efficiency improvements to increase profits (a fact that by itself refutes the notion that actions to reduce greenhouse gases will inevitably hurt the economy). Even administration spokesmen admit that even if the United States achieves the administration's goals on intensity, emissions will rise by 15 percent. Further, even if the U.S. embraced the first steps required by Kyoto, few scientists think that these targets will avert climate change, particularly since it seems to be upon us.

It's not as though the Bush administration ignores all long-term threats; indeed, it has hyped the threat of looming insolvency in Social Security totally out of proportion to estimates of the agency's own actuaries. Rather, the administration looks long term when the threat can be used to advance its short-term agenda (e.g., shifting responsibility for the old-age safety net from the government to individuals).

The United States is not alone in its foot dragging. As a developing nation, China is not bound by Kyoto targets, but that nation's leaders talk a better line on taking action to avert climate change than the U.S. China has higher fuel efficiency standards for automobiles and is looking to environmentally friendly technologies to help meet its enormous appetite for energy. That being said, it's not a stretch for a country to implement high mileage standards when it is already importing most of its oil, and China's push for alternative energy is more than matched by the pace at which it is building coal-fired power plants that will add new carbon to the atmosphere. It's likely that China will surpass U.S. emissions in the not-too-distant future, and India waits in the wings with a billion people.

A simple look at the upward path of global greenhouse-gas emissions answers the question—in the affirmative—of whether we will continue to squeeze the trigger on the gun we have put to our own head. For me, the most depressing aspect of the calamity that we face is the implication that from the perspective of our Martian guest we are no different than fruit flies in our ability to contain our appetites and numbers and to avert predictable calamities. During times of good weather and abundance, we expand to and pass the limits of food and water, and when times turn bad, we crash. For all our vaunted intelligence, our track record suggests that our behavior as a species is ruled by short-term self-interest just like our dim-witted six-legged friends.

Indeed, a good deal of human intellectual firepower seems to be directed at quashing efforts to avert or adapt to climate change. With climate already changing, it's safe to say that we will do virtually nothing to reduce our role in changing climate and will be almost completely unprepared to deal with whatever nasty surprises climate change brings in the future.

While there is resistance to taking action to reduce greenhouse emissions, both industry and government are both more open to programs to either extract carbon from the atmosphere or adapt to climate change once it comes. On the surface, this makes no sense; it is the same as arguing that the best way to deal with someone ingesting arsenic would be to continue to give them arsenic but try to provide them with an antidote as well. From the logic of greed, however, the approach makes perfect sense.

Capitalists resist mandated reductions in greenhouse emissions because (a) they hate regulation in general, and (b) executives feel that actions to reduce emissions will hurt quarterly profits. On the other hand, these same executives love cleanup efforts because they promise future profits. The fractured logic of our market economy first rewards a company for making a mess and then rewards it for providing a solution.

This exactly describes what happened in the 1980s when the world belatedly took action to reduce the emission of chlorofluorocarbons, or CFCs. Studies in the 1970s by atmospheric chemists Sherwood Rowland and Mario Molina showed that these chemicals could harm the ozone layer in the upper atmosphere, a thin shield of oxygen radicals that protects life on earth from lethal UVB radiation. DuPont, the world's largest producer of CFCs successfully lobbied the Reagan administration to halt steps that had been underway since the previous Carter administration. Only when the evidence of damage became overwhelming in the late 1980s did DuPont reverse position, and even then skeptics noted that its motives were suspect. That's because DuPont also had the lead in developing alternatives to CFCs (work begun during the Carter years and then tabled when Reagan was elected), and thus would have a virtual lock on the market if CFCs were phased out. Even better, the profit margins on the alternatives were much higher than on CFCs, which had become a commodity.

This approach of cause-a-crisis-and-then-clean-it-up isn't likely to work with climate change, particularly if it is abrupt. For one thing, the world responded to CFCs before the adverse affects of ozone depletion began imposing costs on the economy (although the biosphere may be paying an ozone-depletion tax in terms of reduced resistance to disease and increases in skin cancer and blindness in the most affected zones). When abrupt climate change makes its presence felt, any profits derived from attempts to adapt to or reverse the changes will be vastly outweighed by the costs of collapsed economies, rampant disease, famines, out-of-control migration, and political upheaval.

Imagine, for instance, a world entering the transition period of a flickering climate that precedes a sudden change from warm to cold. Europe would suffer floods and droughts on a vastly greater scale than those that inundated the continent in 2002 and 2003, and its northern regions would experience intermittent deep freezes as climate and ocean circulation struggled to find a new equilibrium. At the same time, as the jet stream, the ITCZ, ocean currents, ice sheets, sea ice, cloud cover, and other elements of earth's heat distribution system adjusted and reacted to the changing energy equation, droughts and floods not seen since ancient times would afflict some of the most densely populated regions on earth, including China, India, Africa, and Latin America. The probability of drought in the American breadbasket would rise, and along with it the possibility that the American grain surplus—the lion's share of world grain exports—would disappear.

The Christmas tsunami of 2004 in Asia showed both the

vulnerabilities of the modern world to a major natural disaster as well as the resiliency of modern society. During the nineteenth century, a similar tsunami washed through the Indian Ocean. It attracted far less notice, however, in part because fewer people lived right on the beach, there were fewer vehicles and homes to be turned into waterborne projectiles, and there were many more mangroves to absorb the impact. This most recent tsunami produced a heartwarming outpouring of $4 billion in promised (key word) aid, but it also killed roughly 200,000 people. Its derivative effects pushed an estimated 2 million Asians closer to poverty, and set up a situation for even more rapid destruction of Asian forests (timber for reconstruction), reefs (both direct impacts of the wave and the detritus deposited on coral), and mangroves (mostly through damage wrought by waterborne wreckage).

A flickering climate would not have the sudden impact of a tsunami, but it would be affecting virtually every part of the planet at the same time and in so doing reduce the resiliency of the global community. Threshold-crossing events such as the floods in Europe of 2002 would be far more severe and ubiquitous. With every nation dealing with local emergencies, it would be more difficult to mobilize the resources to aid victims in other areas, and there would be fewer resources to mobilize in any event.

Municipalities around the world would struggle under the burden of greatly increased demands on funds to maintain and repair basic infrastructure. FEMA and other safety nets would be bankrupted. The cost of doing business would rise for every conceivable enterprise, reducing profitability as well as the pool of capital. Governments would find tax receipts drastically reduced, and in the world's tightly coupled markets, financial tsunamis would rocket through the system, leaving banks and corporations insolvent. Financial panics, largely absent for over seventy years, would return with a vengeance.

One way to think of the financial impact of a flickering climate is to think of an enormous tax placed on every individual and business. Those lucky enough to have jobs would have greatly reduced incomes. Property values in most places would plummet as buyers disappeared and costs of insurance and maintenance skyrocketed. Even in the developed world, an impoverished population is likely to be an angry population. Another unfortunate lesson of Katrina was that its aftermath exposed how quickly an extreme weather event can reduce a formerly orderly society to anarchy. We may never know the entire tally of crimes committed by gangs in the power vacuum of the first days following the storm, but in a matter of hours, those unfortunates who either couldn't or wouldn't leave New Orleans were plunged into the depraved and dangerous conditions reminiscent of descriptions of the last days of Berlin during World War II when the remaining Germans lay helpless before advancing Russian troops.

A flickering climate would create a world in which there would be enormous opportunities for capital spending but scarce capital to invest. It would be a world with legions of migrants further burdening overburdened governments, massive unemployment, and, very likely, famines and outright starvation. Imagine the horrors of Victorian era El Niños described by Mike Davis for India and China played out on a global scale.

A different kind of tax would fall on the already beleaguered natural world, which in turn would be passed on to humans. As changing climate tilted the board in favor of microbes and other fast-adapting and fast-reproducing pests, pandemics would sweep through animal and plant populations (including staple crops). The combination of drought- and beetle-weakened trees will provide fuel for huge fires, the smoke from which would further depress rainfall. As animals go extinct and habitat disappears, some mammalian diseases would jump over to humans and find fertile incubators in the weakened immune systems of the overstressed poor.

Without doubt, there would be areas of the planet less affected by a flickering climate, but anyplace that promised sanctuary would find itself besieged if not invaded, while many of the worst-hit nations would collapse into civil war. The link between environmental upheaval and instability has been well established—of the fourteen nations that have required U.N. peacekeeping operations since 1990, twelve lost 90 percent of their forests. Add to deforestation such other consequences of climate change as desertification, the loss of agricultural lands and ruined fisheries, and the potential for conflict rises to virtual certainty.

The upper-middle-class American family, today so well protected against external shocks, would find its layers of insulation from the encroaching chaos gradually stripped away. If wage earners still held jobs, they would find themselves without bargaining power, job security, or any cushion against the ever present prospect of financial ruin. Collapsing housing prices would eliminate one cushion of wealth; insolvent banks and recurrent currency panics would breed a fear of the markets. Gold prices would go through the roof, but the family would have to be able to hold onto hard wealth in towns and cities bursting with the seething dispossessed. More likely, families with any savings would stockpile food and necessities in a market economy increasingly prone to breakdowns and interruptions.

Is this the world we really want? Horrific as this vision might be, the reality would probably be far worse if climate goes into one of its tailspins. All I've done is cluster together some events that have happened in recent years. Katrina stretches our capacity to imagine horror, but some of the things that might happen—droughts that last more than a century, an advance of arctic zones southward, incessant and epic storms—simply overwhelm the imagination when we try to apply them to a world of 6 billion people depending on an exquisitely balanced food system.

Is this the world we want—for ourselves, and even more likely for our children? If such a future is inevitable as part of earth's regular climate cycles and we knew it was coming, humanity would probably begin feverish preparations to ride out the coming chaos. The future described above, however, is one we might bring upon ourselves.

One can't be certain that we are creating this hell on earth, but when in the past have we demanded certainty before acting on a potential global threat? Short of certainty, there is a good chance, a better-than-even chance, that some version of the scenario described above lies before us. Yet, like the tourist standing on the beach, we stand passively waiting to see what will happen. It may be that most people are not aware of the risks, particularly given all the qualifications and disinformation that accompany reports on climate change. That excuse gets thinner by the year. If as a nation and as leaders of the world community we continue to refuse to take action to keep this serial killer at bay, or worse invite it into our house, then we will meet our fate.

Chronology:
The Accelerating Pace
of Climate Change and
Scientific Discovery

1950s

In his dissertation, Wallace Broecker publishes his first speculations that climate might be subject to more rapid swings
than was commonly believed by scientists of that time. Scientists at Lamont-Doherty Earth Observatory and elsewhere
find inconclusive evidence of abrupt shifts in temperatures in
the shells of plankton and foraminifera. The dating techniques
of the era, however, had relatively poor resolution, and the
conventional wisdom continues to hew to the belief that climate change is a stately and gradual process.

1960s

Broecker continues to pursue the possibility of rapid climate change. In 1960, along with renowned oceanographer Maurice Ewing, he publishes "Evidence for an Abrupt Change in Climate Close to 11,000 Years Ago," in the *American Journal of Science*. He pursues the notion of rapid change in several other papers in *Science* and other journals.

Other scientists, including Reid Bryson in the United States and Willi Dansgaard from Denmark, report on finding evidence of abrupt climate changes in the analysis of a variety of proxies such as tree rings and ice cores. Willi Dansgaard continues to refine the analysis of ice cores he and his team extract from Greenland and in 1969 publishes in *Science* "One Thousand Centuries of Climate Record from Camp Century in the Greenland Ice Sheet."

1970s

For much of the decade, concerns about climate focus on the prospect of a return to ice age conditions. Paleoclimate evidence suggesting that warm interglacial periods tended to be shorter than previously thought, coupled with evidence of cooling between the 1940s and 1970s, lead a number of scientists, including George Kukla of Lamont-Doherty, Reid Bryson, and Stephen Schneider to speculate that the present warm period might be nearing its end.

By the end of the decade, however, these fears recede and new fears arise that a vast increase in heat-trapping greenhouse gases released by human activities might warm the planet.

1979

• Concerns about the possibility of natural cooling and possible human impact on climate dominate a World Climate Conference in Geneva, Switzerland. Kenneth Hare of the University of Toronto offers the conference declaration: "Climate will continue to vary and to change due to natural causes. The slow cooling trend in parts of the northern hemisphere during the last few decades is similar to others of natural origin in the past, and thus whether it will continue or not is unknown."

• A memorandum prepared for President Jimmy Carter from Gordon McDonald, David Keeling, Roger Revelle, and George Woodwell, four distinguished scientists, addresses the threat of human-caused climate change as a result of greenhouse gas emissions and predicts "a warming that will probably be conspicuous within the next twenty years." The warning is then repeated a year later in the Council on Environmental Quality's *Global 2000 Report to the President*.

The first signs that the global climate was beginning to wobble toward chaos began to garner headlines toward the end of the 1980s.

1988

• From AccuWeather's list of the top events of the twentieth century, "Summer—During the hot summer of 1988, more than 2,000 daily high temperature records were broken. Philadelphia, PA, had a record 18 consecutive 90-degree or higher days. Baltimore, MD, and Washington, DC, both had a record 21 consecutive 90-degree days."

• June 23, James Hansen, an atmospheric chemist at NASA's Goddard Institute for Space Studies in New York, testifies before the Senate Energy Committee during sweltering heat. He asserts: "Global Warming is now sufficiently large that we can ascribe with a high degree of confidence a cause and effect relationship to the greenhouse effect."

• Hartmut Heinrich finds large amounts of iceberg debris in rock cores corresponding to a 7,000–10,000 year cycle. Later these periodic surges of icebergs are identified as key factors in a regular cycle of abrupt climate change and come to be called "Heinrich events" in honor of their discoverer.

• The Office of Polar Programs (OPP) of the National Science Foundation (NSF) officially initiates GISP2, the Greenland Ice Sheet Project to study arctic climate change.

1989

• The United Nations Environment Programme, working with the World Meteorological Society, creates the Intergovernmental Panel on Climate Change (IPCC), convening 2,500 scientists from around the world to examine the links between greenhouse-gas emissions and climate change.

Business pushed back against the growing momentum for an international accord to address the problem of climate change. Lobbyists stressed the uncertainties. Extreme weather events continued, however, and governments began discussing what to do.

1989

• A group of energy companies, automakers, and other industrial giants, including such blue chip companies as DuPont and General Motors form the Global Climate Coalition as a lobbying group to counter efforts to take action on climate change. The GCC operates out of the Washington offices of the National Association of Manufacturers.

1991

• September 16. Typhoon Mireille hits Japan. It is the first typhoon to hit in thirty years and causes $5.4 billion in insured losses.

• October. "The Perfect Storm" hits the coast of New England. At sea, the storm produces hurricane-force winds and 100-foot waves. It hovers off the coast for days, wiping out beaches and homes over hundreds of miles of Atlantic coast.

• Two years into the GISP2 project to extract a two-mile ice core from the center of the Greenland Ice Sheet, the team brings up ice from about one mile below the surface of the sheet, ice that was deposited as snow 11,500 years ago. A few months later, the team extracts cores from the period about 12,800 years ago. The 1,300-year-long cold snap revealed by these cores was first named in the 1930s and is called the Younger Dryas for the arctic flower that flourished during the period. Later, when scientists examine the cores, the scale and rapidity of the changes astonish the scientists.

1991–1992

• Parties meet to develop the United Nations Framework Convention on Climate Change (UNFCCC).

1992

• Under U.S. pressure, delegates to the Rio Earth Summit do nothing to implement the UNFCCC.

• Hurricanes Andrew and Iniki cause $19.5 billion in insured losses. Andrew's $15 billion insured loss ($30 billion total damage) wipes out premium income for the previous twenty years.

1993

• *Nature* publishes Gerard Bond's discovery of signals of rapid climate change in ocean sediments in an article entitled "Correlations in Climate Records Between North Atlantic Sediments and Greenland Ice." Among other things, the paper confirms signals of rapid climate change cycles found in ice cores taken from Greenland.

• "Five hundred year" floods in the Midwest cause $13 billion in damage.

• The hantavirus breaks out in the American Southwest. Epidemiologists tie the outbreak to ENSO, which increases rainfall in the Southwest and spurs a dramatic increase in rodent populations. Mouse densities soar from one per hectare in the summer of 1991 to twenty to thirty per hectare in the spring of 1993.

1994

• Northern India suffers through a "Heat Wave of the Century"—ninety days of temperatures above 100 degrees F. The heat is followed by a dramatic increase in cases of malaria and dengue fever.

• Big European insurance companies such as Munich Re begin to study the impact of climate change. If weather-related calamities increase in the future, insurers will bear the brunt of the losses, since they base their rates on projections of past climate.

• Writing in *Nature,* Broecker identifies massive discharges of icebergs during glacial times as triggers for rapid climate-change events. He gives the name "Bond cycles" to the millennial-scale pattern of abruptly changing temperatures that follow these discharges.

1995

• January. The Larsen A Ice Shelf, stable for hundreds of years, disintegrates, losing 1,700 square kilometers in a single week.

• "Flood of the Century" in the American Southeast.

• Warmest year on record.

• Writing in *Science,* Bond further elucidates the connection between iceberg discharges and rapid cooling and warming, and shows that such cycles can occur during warm periods as well as glacial times. Speaking at the fall meeting of the American Geophysical Union, Bond says, "Because we now think that climate flips can occur on the earth relatively free of ice, the odds of a future climate jolt could be higher than we thought."

1997

• "Flood of the Century" inundates the Red River Valley in the upper Midwest.

• Second-hottest year on record.

• Scientists find increasing evidence of abrupt shifts in climate during the Holocene, a period previously thought to be a period of climate tranquillity. Typical of these papers is one by Peter deMenocal and Gerard Bond entitled "Holocene Climate Less Stable than Previously Thought," published by the American Geophysical Union in its *EOS Transactions.*

• *Science,* October 31, K. C. Taylor et al. publish "The Holocene–Younger Dryas Transition Recorded at Summit, Greenland." The article presents an analysis of the end of the Younger Dryas in view of the ice core data first uncovered by Alley and colleagues in 1991. The article presents the possibility that climate change may not be incremental and moderate, as the IPCC assumes, but rapid and large.

• In *Science,* November 28, Wallace Broecker publishes "Thermohaline Circulation, the Achilles Heel of Our Climate System: Will Man-Made CO_2 Upset the Current Balance?" The article links the abrupt changes in climate at the beginning and end of the Younger Dryas to changes in ocean circulation and explicates how global warming can lead to episodes of rapid cooling. In his conclusion he writes: "Through the record kept in Greenland ice, a disturbing characteristic of the Earth's climate system has been revealed, that is, its capability to undergo abrupt switches to very different states of operation. I say 'disturbing' because there is surely a possibility that the ongoing buildup of greenhouse gases might trigger yet another of these ocean reorganizations and thereby the associated large atmospheric changes. Should this occur when 11-to-16 billion people occupy our planet, it could lead to widespread starvation, for in order to feed these masses, it will be necessary to produce two to three times as much food per acre of arable land than we now do. More problematic perhaps than adapting to the new global climate produced by such a reorganization will be the flickers in climate that will likely punctuate the several-decade-long transition period."

• The Strongest El Niño in 130,000 years (as evident from records encoded in ancient coral) causes $100 billion damage worldwide, leaving in its wake catastrophic floods along the Yangtze in China, massive fires in Indonesian Borneo and the Brazilian Amazon, and extreme drought in Mexico and Central America.

• The Kyoto Treaty to control greenhouse-gas emissions is finalized in Japan and sent to member states for ratification. It contains no mandatory emissions targets for developing nations, including China, the most populous country and the fastest-growing economy on the planet.

• BP drops out of the Global Climate Coalition.

1998

• January. A massive ice storm in upstate New York and Canada leaves millions without power for days.

• The year replaces 1995 as the warmest year on record.

• October 2. Hurricane Mitch hits Central America with 180 mph winds. By the time it loses its punch on November 5, it has killed 10,000 people, mostly through floods and mudslides.

• El Niño continues to raise havoc with weather around the world. "Weather-related losses of $89 billion for the first 10 months of the year exceed all weather-related losses for entire decade of 1980s." (heatisonline.org) On the heels of the Asian economic meltdown of 1997, El Niño–related drought stunts the rice crop in Indonesia. Resulting food riots and ethnic strife bring down the Suharto government.

• *Nature* publishes a study led by Michael Mann of the National Climatic Data Center which argues that temperatures have risen sharply in the latter part of the twentieth century, making the period demonstrably warmer than any during the past several hundred years. A year later, the distinctive pattern shown by this data comes to be referred to as the "hockey stick."

• The United States suffers a record-setting 1,424 tornadoes.

1999

• From a NOAA press release: ". . . U.S. EXPERIENCES SECOND WARMEST YEAR ON RECORD; GLOBAL TEMPERATURES CONTINUE WARMING TREND. . . . Global temperatures for 1999 are expected to be the fifth warmest on record since 1880, NOAA and the World Meteorological Organization reported."

• Hurricanes Dennis, Floyd, and Irene hit North Carolina in a matter of weeks, the first time in over fifty years that the state has been hit by three hurricanes in a single season. Floyd drops 15 inches of rain on already saturated land. Three hundred towns are completely submerged, 30,000 homes damaged and 10,000 destroyed, along with roads, bridges, dams, and community drinking supplies. Total losses exceed $5 billion.

2000

By 2000, signs of changing climate have become so obvious and the scientific consensus so solid that a number of major corporations decide that it is ill advised to belittle the threat. Still, the election of George W. Bush reduces the impetus for U.S. government action on the problem.

• Flooding in Mozambique cuts GDP by 45 percent, leaves 1 million homeless.
• Following the defections of American Electric Power, Dow, DuPont, Royal Dutch Shell, Ford, DaimlerChrysler, Southern Company, Texaco, and General Motors, companies concerned about appearing cavalier about the threat of climate change, the GCC restructures around trade associations with the hope that corporations can continue to contribute but still protect their image.
• Associated Press, March 12: ". . . the last three winters have been the warmest on record in the United States—a pattern of warm winters established in 1980, said the scientists. Since then, 67 percent of the winter seasons have been warmer than the long-term average."
• The year goes into the books as the fifth-warmest on record.

2001

• A "Hundred-Year Drought" continues in South Florida, triggers 1,200 wildfires and causes 4 billion in damage throughout the state.

• June. Tropical Storm Allison becomes the costliest tropical storm in history as it inflicts $4 billion in damage from Louisiana and Texas to Pennsylvania. The damage comes not from winds (which never topped 60 mph), but rain: 18 inches in Baton Rouge, 12 inches in Houston, 9 inches in Bucks County, Pa. For the Federal Emergency Management Agency, the storm represents the first $1 billion loss in the history of its Flood Insurance Program. Largely because of Allison, the Flood Insurance Program ends up owing the U.S. Treasury $560 million.

• The year replaces 1997 as the second-hottest on record.

• September 11. The destruction of the World Trade Center pushes concerns about climate to the background. Swiss Re sustains a loss of 3 billion Swiss francs in the tragedy. Swiss Re and other companies realize that they had previously assumed the risk of terrorist attack for free and review other risks, such as climate change, that they have not passed on to customers.

2002

• The year replaces 2001 as the second-hottest on record.

• Swiss Re hosts a conference at the American Museum of Natural History in New York City. New York State and other officials describe potential lawsuits they might bring against companies that contribute heavily to greenhouse-gas emissions.

• August 14. Church World Service, Situation Report: "The worst floods in 500 years in eastern and central Europe have been rapidly spreading across the continent this week, causing severe damage to housing and public infrastructure and forcing thousands of people to flee their homes.

"Worst affected by the flooding is the Czech Republic: some 200,000 have fled the capital of Prague and southern Bohemia. Floods have also caused severe damage in the Black Sea region of the Russian Federation, southern parts of Romania, Austria, Germany, Hungary, Slovakia and Bulgaria.

"Today floodwaters advanced on Dresden, Germany, where army helicopters flew thousands of Dresden hospital patients out of the reach of the rising water, Reuters reported."

• NOAA issues a statement noting that by September the arctic ice cap had shrunk to the smallest area of any period during twenty-four years of satellite observations.

• With few paying members, the Global Climate Coalition disbands, defiantly noting in their final statement that the Bush administration has adopted their approach to climate change.

• By the end of the year, seventeen states have begun developing their own policies to control greenhouse gases.

• In *Science,* "The Perfect Ocean for Drought," Martin Hoerling and Arun Kumar report that an analysis of sea-surface temperatures from 1998 to 2002 reveals a different post–El Niño pattern than any previously recorded in the century, a pattern characterized by an unprecedented warming of the western Pacific and a persistent cold state in the eastern ocean. The authors argue that the reduced evaporation from the unusually cool eastern tropical Pacific explains the persistent drought in the mid-latitudes, including much of the western United States. The authors argue that warming in this century accounts for the abnormal oceanic conditions.

• FEMA, which insures 4.4 million flood-prone properties, takes a series of actions to mitigate ever costlier natural disasters. These include eliminating subsidies for second homes and raising premiums to include the cost of anticipated erosion-related losses. Realty organizations decry the changes.

2003

• Associated Press, February 28: ". . . The world has experienced unusually extreme weather in recent decades and economic losses from storms and other catastrophes have increased tenfold, an independent research group reported Thursday. . . . the council said there were 26 'major flood disasters' worldwide in the 1990s, compared to 18 in the1980s, eight in the 1970s, seven in the 1960s and six in the 1950s. The largest number of severe floods occurred in Asia, the council said."

• In *Science,* March 28, R. B. Alley et al. publish "Abrupt Climate Change," which argues that past episodes of rapid climate change have been more frequent than previously thought, and that even small-scale events of rapid change in modern times have resulted in disruption of economies, agriculture and fisheries."

• April 20. Tropical Storm Ana forms 250 miles west of Bermuda. Ana is the earliest Atlantic tropical storm on record.

• Cold, snowy weather persists in the U.S. Northeast until May. A monster President's Day snowstorm in February paralyzes the east coast from Washington to Boston (where the 27.5 inches of snow was the largest total on record). By the end of the season, much of the east coast had received twice the average annual snowfall.

• John Dutton, dean emeritus of Penn State University's College of Earth and Mineral Sciences, estimates that $2.7 trillion of the $10 trillion U.S. economy is susceptible to weather-related loss of revenue (with many trillions more at risk in the form of buildings vulnerable to storms and floods).

• Writing in the social science journal *Energy and Envi-*

ronment, Stephen McIntyre, a mining executive and Ross McKitrick, an economist, challenge the "hockey stick," arguing that the mathematical methods Mann used to present his data have a bias toward generating hockey stick–looking graphs. While *Nature* rejects a submission by McIntyre and McKitrick, their contention is taken up by climate change skeptics.

• July–August. Record-setting heat kills more than 35,000 people in Europe. Temperatures reach 114 degrees F. at Cordoba Airport in Spain. Crop harvests are reduced by 5 percent across Europe, and the value of crop losses in Germany alone surpasses 1 billion euros. Wildfires sweep through France, Spain, and Portugal.

• September. Heat and drought in Europe so reduce the flow to the Rhine that shipping is interrupted. At Kaub, Germany, the river drops to 13.8 inches, an all-time low and nearly 7 feet lower than the average seasonal depth at this point.

While governments in Europe and Japan chafe for action on climate change, the U.S. continues to balk. Despite the ever growing tally of extreme events, some in government and the media continue to dismiss the threat.

• Swiss Re, the world's largest issuer of so-called D&O insurance (policies that protect directors and officers of corporations from shareholder lawsuits), announces that it is sending out questionnaires to all its policyholders asking about company positions on climate change. The move is a precaution prompted by the possibility of international action and subsequent shareholder suits against companies that through their actions have contributed significantly to greenhouse emissions and failed to take any mitigating measures. Should the international community set limits on greenhouse emissions, Swiss

Re has set the stage to exclude climate-change-related lawsuits from these insurance policies. This would leave the affected executives and directors personally liable for damages resulting from such suits.

• Brush fires in Southern California destroy hundreds of homes. Fires spread through pine trees weakened by beetle infestations and dried out by several years of drought. The fires damage 743,000 acres of brush and timber and burn 3,700 homes.

• December. One hundred twenty nations accounting for 44.2 percent of emissions by the world's industrial countries have ratified the Kyoto Protocol.

• *New York Times,* December 3, front page: "Russia to Reject Pact on Climate, Putin Aide Says; Kyoto Accord in Trouble; Without Moscow, Treaty to Curb Harmful Gases Cannot Take Effect." ". . . A senior Kremlin official [Andrei N. Illarionov] declared Tuesday that Russia would not ratify the international treaty requiring cuts in the emissions of gases linked to global warming, delivering what could be a fatal blow to years of diplomatic efforts. . . . 'A number of questions have been raised about the link between carbon dioxide and climate change which do not appear convincing,' Mr. Illarionov said. . . ."

• *Wall Street Journal* (lead editorial), December 4: "President Bush could kill two shibboleths with one stone were he to pick up the phone and tell Vladimir Putin what an asset the Russian President has in Andrei Illarionov. . . . Mr. Illarionov in particular has repeatedly had the courage to say what most politicians won't: both the science and the economics of Kyoto are fundamentally flawed. . . . There's also the nagging problem that temperatures more than 1,000 years ago at times appear to have been as warm, if not warmer, than today's."

• Same day, *Wall Street Journal,* page A3: "This Year

Likely to Be Third Hottest; Warm Fall Is Cited; Many Scientists Say Trend to Higher Temperatures Is Due to Gas Emissions." "Meteorologists say 2003 is on its way to being the third-hottest year since modern temperature readings began. . . . The record-setting heat marks the 27th consecutive year that average temperatures have exceeded historical averages. . . . 2001 is now expected to slip to fourth place. . . . Consistent thermometer readings began only in the mid-19th century, but scientists have extracted temperature records from tree rings and buried ice going back several millennia. Those data suggest that recent temperatures may be higher than any over the past 2,000 years.

"Scientists say the pace of change is also increasing, with temperatures rising three times as fast during the past three decades than over the entire 20th century."

• *Science,* December 5, publishes "Modern Global Climate Change," by Thomas Karl of the National Climatic Data Center and Kevin Trenberth of the National Center for Atmospheric Research. The two scientists argue that "modern climate change is dominated by human influences . . ." and that "we are now venturing into the unknown with climate, and its associated impacts could be quite disruptive."

• December 7. Tropical Storm Odette dissipates after killing two people in the Dominican Republic. Odette is the first named tropical storm to hit the Caribbean in December. It formed on November 27, four days after the official end of the Atlantic hurricane season.

• Same day. A monster snowstorm lashes a 750-mile swath of the east coast with up to 52 inches of snow. Winds off Maine reach Class 1 hurricane force—82 mph. New York is buried under 20 inches of snow. Records fall across the East for the largest snowfall in early December. Mayor Bloomberg notes that it costs New York $1 million an inch to clear the city's streets of snow. Two weeks before the official start of winter, much of the east coast had received between 25 and 40 percent of a full winter's snowfall.

• December 9: Tropical Storm Peter forms in the eastern Atlantic. According to the National Hurricane Center, this marked the first time since 1887 that two tropical storms had formed in December. The year thus goes into the record books as having witnessed the earliest and latest tropical storm activity on record.

• Because of continued drought, competition for water, and other factors, China's grain production continues to decline from peak production of 512 metric tons in 1998. By 2003, the shortfall between production and consumption amounts to 45 million metric tons—the equivalent of Canada's total production.

2004

• A consortium of scientific organizations publishes the Arctic Climate Impact Assessment. A compilation of scientific studies conducted in the Arctic, the assessment addresses the threats of further warming, noting that "arctic average temperatures have risen at almost twice the rate of the rest of the world over the last few decades."

• In the December meeting of the American Geophysical Union, Caspar Ammann, a paleoclimatologist at the National Center for Atmospheric Research and Eugene Wahl of Alfred University refute McIntyre and McKitrick's assertion that the data don't show a "hockey stick" rise in temperatures in the latter half of the twentieth century. Their methods produce much the same results that Mann and colleagues produced in 1998.

• The *New York Times*, December 12, reports that shrinking mountain snows have exacerbated a seven-year drought in Afghanistan. Rivers no longer reach their basin in Nimruz, and the region's rainfall for the past three years has averaged one-twentieth their meager normal levels of 2.3 inches.

• By the end of September, 2004 has already broken the record for tornadoes in the United States. By the end of the year, 1,817 tornadoes have been logged.

• Hurricanes Charley, Francis, Ivan, and Jeanne devastate parts of Florida and the Caribbean, and also contribute to the record-breaking number of tornadoes. Insurance claims resulting from storms total $43 billion.

• The World Meteorological Organization ranks the year as the fourth-hottest on record.

2005

• A series of storms from late December 2004 to early January 2005 deposits up to 19 feet of snow in the Sierra Nevada—the most recorded since 1916.

• Following the record-breaking year in 2004, Tornado Alley in the United States is unusually quiet in 2005. While Oklahoma normally has ten tornadoes in April and twenty in May, in 2005 the state had only one in April and none in May.

• May is the third-coldest on record in the Northeast. Late in the month, a northeaster batters the Massachusetts and Maine coasts for nearly a week with winds up to 60 mph.

• On May 17, Adrian forms in the Pacific and becomes the earliest hurricane to cross Central America this early in the season since the National Hurricane Center began keeping records.

• On June 8, Andrew Revkin reports in the *New York Times* that documents obtained by a former government official, Rick Piltz, reveal that Philip Cooney, a lawyer with no scientific training, edited and weakened scientific assertions in a government report on climate change while working for the White House Council on Environmental Quality. Before joining the government, Cooney had helped lead attempts by the

American Petroleum Institute to prevent the United States from taking action to prevent climate change.

• On June 11, Cooney resigned his post. Four days later, Cooney took a job with ExxonMobil, the world's largest oil company.

• In the spring and summer, several more analyses refute the challenge to Mann's "hockey stick" chart of past temperatures. Despite this, in July, Representative Joe Barton, a Texas Republican on the House Appropriations Committee, subpoenas personal records as well as scientific data from Mann and other scientists associated with the hockey stick.

• June and July break records for heat throughout the northeast United States. The extremely long heat wave in the east is blamed on an unusually persistent Bermuda high. The American west suffers under its own heat wave as temperatures approach 120 F degrees in Phoenix and Las Vegas. Scores of homeless and shut-ins die in the region as a result of the heat.

• August 29. Hurricane Katrina hits the Gulf Coast and destroys a major American city. The storm is the worst natural disaster in American history.

• September 28. Republican Senator James Inhofe, chairman of the committee on Environment and Public Works, holds hearings on climate change and turns to Michael Crichton, a science fiction writer, for expert opinion. Crichton criticizes climate scientists for not performing randomized double-blind studies. In response, David Sandalow of the Brookings Institution notes that when it comes to planet earth there is only one experimental subject.

• October. Hurricane Wilma forms and in a matter of hours becomes the most intense Atlantic hurricane ever. Following Wilma, Hurricane Alpha breaks the previous record for number of named storms and is quickly followed by Beta. As of this writing, new storms threaten, and 2005 will likely be the warmest year on record.

Acknowledgments

Thanks to Alice Mayhew who tolerated what turned out to be a longer trip than either of us anticipated.

To the Mesa Refuge, which provided a retreat at the beginning of the writing, and

To my friends Ani and Jerry Moss, who provided one at the end.

To my late friend Leon Levy, who offered encouragement along the way.

To Esther Newberg for her loyalty and sensible advice.

To Mary and the kids for putting up with a writer's eccentricities.

To the many institutions that facilitated my research during the several years this book was in process, notably, the National Science Foundation, Lamont-Doherty Earth Observatory, and Scripps Institution of Oceanography. Special thanks

Acknowledgments

to Woods Hole Oceanographic Institution for allowing me to accompany a research mission across the Gulf Stream.

To Eliza Little and David Bjerklie for their excellent research and fact checking. And, finally, thanks to the many, many scientists who helped me understand the workings of climate and how it changes during the seventeen years I have been following this issue. If I started singling out individuals, many forests would have to fall to accommodate all the names, and I'm fairly certain the scientists in question would prefer that the trees continue to temper climate and sequester carbon.

Index

About the Author

Eugene Linden writes about the environment, nature, science, and technology. He has written for *Time*, including numerous cover stories, for almost twenty years, and has contributed articles and essays to *The Atlantic, National Geographic, The New York Times Magazine, Foreign Affairs, Condé Nast Traveler, The New York Times, The Wall Street Journal,* the *Los Angeles Times, Fortune,* and *Slate.* He is the author of seven books, including *The Octopus and the Orangutan: More Tales of Animal Intrigue, Intelligence, and Ingenuity; The Parrot's Lament: And Other True Tales of Animal Intrigue, Intelligence, and Ingenuity;* and *The Future in Plain Sight,* which the *Rocky Mountain News* called "the most important book of the decade." Linden speaks frequently about nature and the environment and is the recipient of numerous journalism awards. He lives in Washington, D.C.